ADVANCED MATHS FOR AQA

Decision Maths

Brian Jefferson

Course consultant: Geoff Silcock

D1

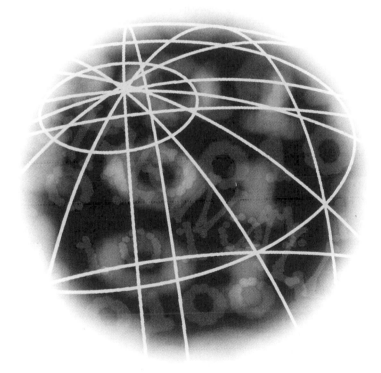

OXFORD
UNIVERSITY PRESS

OXFORD

UNIVERSITY PRESS

Great Clarendon Street, Oxford OX2 6DP

Oxford University Press is a department of the University of Oxford.
It furthers the University's objective of excellence in research, scholarship,
and education by publishing worldwide in

Oxford New York

Auckland Bangkok Buenos Aires Cape Town Chennai
Dar es Salaam Delhi Hong Kong Istanbul Karachi Kolkata
Kuala Lumpur Madrid Melbourne Mexico City Mumbai Nairobi
São Paulo Shanghai Taipei Tokyo Toronto

Oxford is a registered trade mark of Oxford University Press
in the UK and in certain other countries

British Library Cataloguing in Publication Data

Data available

ISBN 0 19 914988 7

10 9 8 7 6 5 4 3 2 1

Typeset by Tech-Set Ltd, Gateshead, Tyne and Wear
Printed and bound in Great Britain by Bell and Bain.

Acknowledgements
The publishers would like to thank AQA for their kind permission to reproduce
past paper questions. AQA accept no responsibility for the answers to the past
paper questions which are the sole responsibility of the publishers.

The publishers would also like to thank James Nicholson for his authoritative
guidance in preparing this book.

The image on the cover is reproduced courtesy of Royalty-Free/Corbis.

About this book

This advanced level book is designed to help you to get your best possible grade in the AQA D1 module for first examination in 2005. This module can contribute to an award in GCE AS-level Mathematics or A-level Mathematics.

Each chapter starts with an overview of what you are going to learn.

This chapter will show you how to

◆ Model matching problems as bipartite graphs
◆ Find an alternating path through a partially matched graph
◆ Use the maximum matching algorithm to improve on a known matching or to establish that the known matching is maximal

Key information is highlighted in the text so you can see the facts you need to learn.

> The number of odd vertices in any graph is even.

Worked examples showing the key skills and techniques you need to develop are shown in boxes. Also marginal hint boxes show tips and reminders you may find useful.

Example 2

Arrange the numbers 4, 8, 2, 6, 3, 5 in ascending order, using the shuttle sort algorithm.

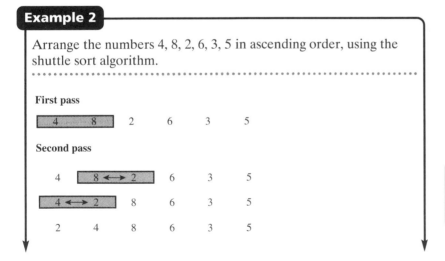

First pass

| 4 | 8 | 2 | 6 | 3 | 5 |

Second pass

| 4 | 8 ←→ 2 | 6 | 3 | 5 |

| 4 ←→ 2 | 8 | 6 | 3 | 5 |

| 2 | 4 | 8 | 6 | 3 | 5 |

> $8 > 2$, so swap.
> Compare first and new second numbers: $4 > 2$, so swap.

1 Introductory ideas

This chapter will show you how to

- Define an algorithm
- Appreciate the need for an algorithmic approach to certain types of problem
- Appreciate that a graph or a network can be used to model relationships between items in a variety of situations
- Apply some of the terminology of graph theory
- Represent a graph or a network by means of an adjacency matrix or a distance matrix

1.1 What is decision mathematics?

Decision mathematics covers a number of techniques which are important in the solution of large-scale organisational and business problems. Much of the underlying theory has been around for some time, but its range of application has been greatly widened by the development of computers.

Fundamental to decision mathematics is the notion of an **algorithm**, which is a set of instructions – a sort of recipe – which can be blindly followed to obtain the solution to problems of a given type. This chapter contains an informal introduction to the idea of an algorithm. A more formal treatment can be found in Chapter 9.

Many of the situations you will need to analyse can be modelled by drawing a diagram in which items are represented by points and the relationship between the items by lines connecting the points. Such diagrams are called **graphs** or **networks**. An introduction to some of the basic theory and terminology is included in the present chapter. Various uses of these ideas are covered in Chapters 2 to 6, and more graph theory is introduced as it is needed.

The London Underground map is a graph.

Chapters 7, 8 and 9 are, to a large extent, distinct and can be tackled at any stage.

What is an algorithm?

An algorithm is a well-defined sequence of steps leading to the solution of problems of a given type.

To get a feel for the notion of an algorithm, consider this simple puzzle. It involves six counters – three black and three white – placed on a grid of seven squares, as shown.

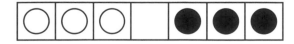

The white counters may only be moved from left to right, either sliding into a vacant space or jumping over one black counter into a

D1

vacant space. The black counters may be moved right to left under the same rules. The aim, of course, is to reverse the positions of the two sets of counters.

If you are not familiar with the solution to this puzzle, try it before reading on.

Communicating the solution

When writing down the solution to the puzzle, you could just list all 15 moves needed. However, this is unnecessarily lengthy. It is also limiting, because it is of no help in related situations, such as four counters of each colour.

It is better to write a list of instructions which will enable anyone to decide on the right move at each stage. One possible list is:

Step 1 Decide at random which colour is the 'active' colour.
Step 2 Move the front active counter.
Step 3 If no further active-colour move is possible, go to Step 6.
Step 4 If the next available active-colour move is a **second** slide move, do not make the move but go to Step 6.
Step 5 Make the next available active-colour move and go to Step 3.
Step 6 If the puzzle is complete, then stop.
Step 7 Change the active colour and go to Step 2.

> Try following these instructions and see that they do indeed lead to the solution.

These instructions are more general than are necessary to cover just the solution of the original puzzle. They will work for any number of counters of each colour, even if there is not the same number of each colour. **Generalising** is an important feature of all algorithms.

An algorithm should have the following features:

✦ It should enable anyone to solve all problems of a particular type, just by following the instructions. No insight into the problem should be needed.

✦ It should provide a clear 'next step' at each stage of the solution.

✦ The problem should be solvable in a finite and predictable number of steps. In the example with counters, m white counters and n black counters would require $(mn + m + n)$ moves – you might try proving this.

> It is important to realise that questions on this topic are not testing whether you can find the solution but whether you know and can apply the correct algorithm. You should therefore always follow the **correct steps** and make your working **clear**.

Most real-world problems are too large to make manual solution viable. Because of this, another desirable feature of an algorithm is that it should readily lend itself to implementation on a computer.

Exercise 1A

1 A ferry company has a small vehicle ferry with four lanes, each 40 m long. The algorithm they use for positioning vehicles in the lanes is as follows.

>   **Step 1** Examine the next vehicle.
>
>   **Step 2** If the vehicle will fit into the first lane, place it there and go to Step 1.
>
>   **Step 3** If the vehicle will fit into the next lane, place it there and go to Step 1.
>
>   **Step 4** If untried lanes remain, go to Step 3.
>
>   **Step 5** Stop.

a) The spaces, in metres, needed for the vehicles on a certain day (in order of arrival) were as follows.

>   4, 6, 5, 12, 14, 5, 5, 6, 4, 4, 14, 16, 10, 5, 6, 5, 8, 13, 6, 4, 4

Use the algorithm to place them in the lanes and record what you find.

b) It is then suggested that they amend their approach as follows.

As vehicles arrive, line them up in two queues, one for vehicles 10 m or over, the other for shorter vehicles. Load first the long vehicles, then the short vehicles, using the previous algorithm.

What would be the result of using this modified approach for the vehicles listed above?

2 A large group of adults travelling in a remote region found their route blocked by a river. They met two children who owned a boat, but the boat was only big enough to carry the two children or one adult. Nevertheless, the children managed to ferry the party across the river and return to their starting point.

a) Give a list of instructions to explain, as economically as possible, how this was done.

b) Find how many times the boat had to cross the river if there were ten adults in the party.

> Neither of the algorithms in Question **1** guarantees to find the best (optimal) solution to a particular case. They are both **heuristic algorithms** – good but not necessarily optimal solutions to problems in a reasonably quick time. No known algorithm guarantees to find the optimal solution every time to packing problems of this sort

D1

> In the examination, you will not be expected to write algorithms from scratch, although you may be asked to modify or complete given algorithms.

1.2 Basic ideas of graphs and networks

In this context, a **graph** refers, not to the familiar axes and coordinates, but to a diagram involving a set of points and interconnecting lines.

Graphs

Graphs have become important because they can be used to model a wide variety of problems. Examples are problems involving:

✦ Places and journeys

✦ Components and electrical connections

✦ People and relationships

✦ Atoms and chemical bonds

✦ Species and predator/prey relationships in an ecosystem

The common feature of these and many other situations is that they involve items (represented by the points of the graph) and relationships (represented by the connecting lines).

> Each point on a graph is called a **vertex** (plural **vertices**)
> Each connecting line is called an **edge**

The terms vertex and edge come about because graph theory has its roots in the study of solid shapes, the points and lines corresponding to the corners (vertices) and edges of the solid. Some authors and syllabuses use the alternative terminology **node** and **arc** in place of vertex and edge.

Vertex or node

Edge or arc

It is important to realise that it is the **way in which vertices are or are not connected to each other** which matters. The shape or layout of the diagram is irrelevant (although obviously a tidy layout helps you visualise the situation). The three diagrams below show the same set of vertices and edges and could model the same situation. They are effectively the same graph.

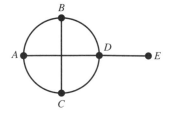

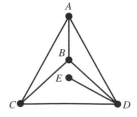

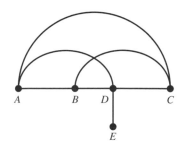

You should check that you can see how these are essentially the same.

A graph is **connected** if it is possible to travel from any vertex to any other vertex (perhaps passing through other vertices on the way).

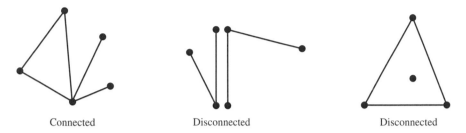

Connected   Disconnected   Disconnected

In some graphs, there are two or more edges connecting the same pair of vertices. There may also be a loop connecting a vertex to itself.

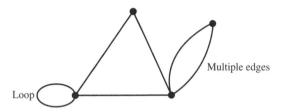

Loop   Multiple edges

A graph with no loops or multiple edges is a **simple graph**.
A simple graph is **complete** if there is an edge connecting every possible pair of vertices.

Complete graphs are important enough to have their own notation:

K_n is the complete graph with n vertices.

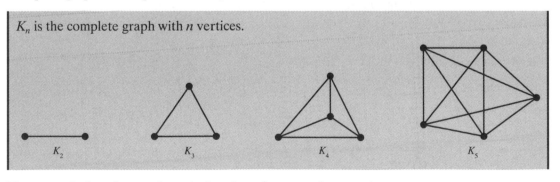

K_2 K_3 K_4 K_5

You can calculate the number of edges in a given complete graph as follows. Consider the complete graph K_6. Each vertex is connected to the other five and so has five edges attached to it. This would give $6 \times 5 = 30$ edges. However, each edge would have been counted twice (once at each end), so the actual number of edges is 15.

Similarly, K_7 has $\frac{1}{2} \times 7 \times 6 = 21$ edges.

In general, K_n has $\frac{1}{2}n(n - 1)$ edges.

The degree of a vertex

If you count the number of edges arriving at a vertex, you have the
degree of the vertex (sometimes called the **order** or **valency**). In the
example (right), each vertex has been labelled with its degree.

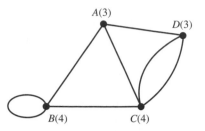

Example 1

Draw a graph with four vertices – one with degree 4, one with
degree 2 and two with degree 1.

Here are two possible graphs which fit the facts.

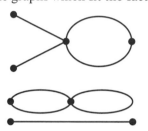

There is at least one other
connected solution and one
other disconnected solution. You
might like to try to find these.

Exercise 1B

1 In each case, draw a connected graph with the given information.
  a) One vertex of degree 3 and three vertices of degree 1.
  b) One vertex of degree 1 and three vertices of degree 3.
  c) Two vertices of degree 3 and two vertices of degree 1.
  d) Four vertices, each with a different degree.

2 In each of these sets of four diagrams, three represent the same
graph and the fourth is different.

Identify the odd one out.

a) i)   ii)    iii)    iv)

b) i)   ii)   iii) 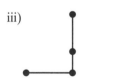   iv)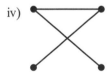

c) i) 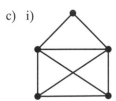   ii) 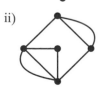   iii) 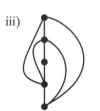   iv)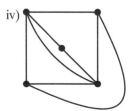

1.3 Representing graphs as matrices

You have seen that the same graph can be drawn in many ways and that it can be difficult to see that two diagrams show the same graph. However, you can store the information of a graph by means of a table or **matrix**.

> The plural of matrix is **matrices**.

If there is at least one edge connecting two vertices, the vertices are said to be **adjacent** and as a result the matrix is called an **adjacency matrix**. It records the number of direct routes from one vertex to the other. (If there is a loop at a vertex, it counts as two routes between the vertex and itself). Here is an example.

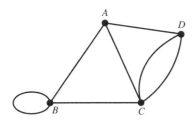
$$\begin{array}{c@{}c} & \begin{array}{cccc} A & B & C & D \end{array} \\ \begin{array}{c} A \\ B \\ C \\ D \end{array} & \left[\begin{array}{cccc} 0 & 1 & 1 & 1 \\ 1 & 2 & 1 & 0 \\ 1 & 1 & 0 & 2 \\ 1 & 0 & 2 & 0 \end{array}\right] \end{array}$$

> This graph has five pairs of adjacent vertices: A, D; A, C; A, B; B, C and C, D.

D1

Notice that if you add up the column (or row) for a given vertex, it gives the degree of that vertex.

Networks

A network is a graph with the added feature that each edge has an associated number, called a **weight**.

You can represent a network by means of a table or matrix, called a **distance matrix**. In this case, the entries in the table correspond to the weights of the edges. For example, here is a network showing average travelling times (in minutes) between towns, and the corresponding matrix.

> The weight may represent a distance, a time, a cost or many other things, depending on the nature of the situation being modelled.

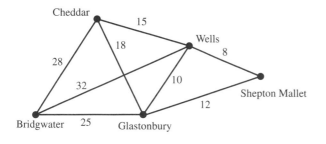

From \ To	Bridgwater	Cheddar	Glastonbury	Shepton Mallet	Wells
Bridgwater	–	28	25	–	32
Cheddar	28	–	18	–	15
Glastonbury	25	18	–	12	10
Shepton Mallet	–	–	12	–	8
Wells	32	15	10	8	–

Some graphs have no cycles. An example is shown on the right.

A graph like this is called a **tree**.

You will have met such graphs in diagrams of family trees, or in drawing tree diagrams for probability.

2.2 Spanning trees

The network diagram below shows a country park, with seven picnic sites and a car park joined by rough paths. Distances are in metres. It is planned to lay tarmac on some of the paths, so that wheelchair users can reach all the picnic sites. Obviously, it is desirable to keep the cost to a minimum, so the choice of paths on which to lay tarmac is important.

D1

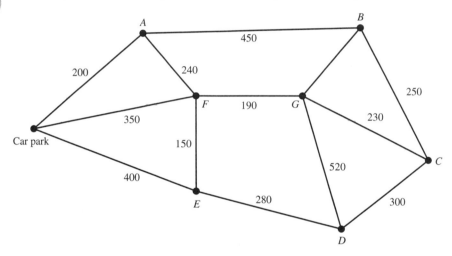

The shortest paths possible must be used, and no paths must be laid with tarmac unnecessarily. The best solution is given below.

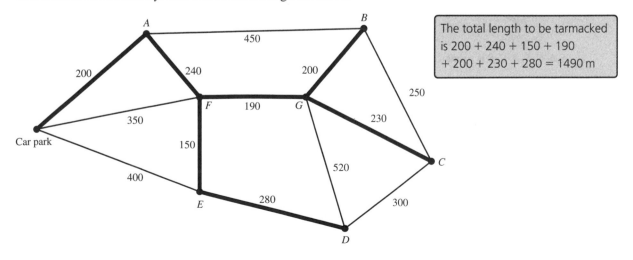

The total length to be tarmacked
is 200 + 240 + 150 + 190
+ 200 + 230 + 280 = 1490 m

The edges which have been chosen, shown as thicker lines, form a tree. Any additional edge would form a cycle and is unnecessary, because it is possible to travel from any vertex to any other by using only the edges in the tree.

The tree is called a **connector** or a **spanning tree** (because it is a tree that spans the whole network). This particular tree is a **minimum connector** or **minimum spanning tree** because the total of the weights is as small as possible.

> A **connector** or **spanning tree** is a tree that includes all the vertices of a graph.
> A **minimum connector** or **minimum spanning tree** is the connector or spanning tree with the least total weight.

The problem of finding a minimum connector is an important one, with applications in the field of transport and distribution planning, designing the layout of computer networks or installing services such as telephone and TV cables.

D1

Although the previous example was of a sufficiently small scale for the solution to be fairly obvious by inspection, the number of possible trees increases very rapidly as the complexity of the network increases. You need a systematic algorithm to enable you to solve more complex problems.

Kruskal's algorithm

The first algorithm was developed by Martin Kruskal. It is an example of a **greedy algorithm**, because at each stage you make the most obviously advantageous choice without thinking ahead. As it turns out, this greedy approach always leads to the minimum connector.

Kruskal's algorithm is as follows:

Step 1 Choose the edge with the minimum weight

Step 2 Choose from the remaining edges the one with the minimum weight, provided that it does not form a cycle with those edges already chosen.

Step 3 If any vertices remain unconnected, go to Step 2

If at any stage there is a choice of edges, choose at random.

You will know when you have finished, either by checking that there are no more unconnected vertices, or by counting the number of edges that you have chosen.

> When a network has n vertices, it will need $(n - 1)$ edges to form a spanning tree.

You can easily show that this is so, by realising that a network with two vertices is spanned by one edge, and that each extra vertex requires one extra edge.

Example 1

List the order in which Kruskal's algorithm would choose the edges in the country park described on page 10.

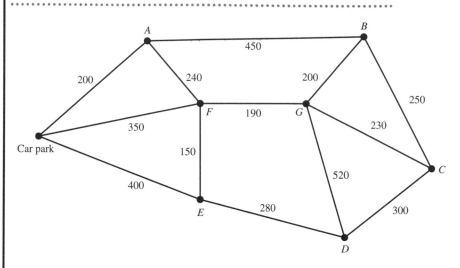

1st	*EF*	150	(*EF* has the lowest weight)
2nd	*FG*	190	
3rd	*BG*	200	
4th	*Car park – A*	200	(3rd and 4th choice in random order)
5th	*GC*	230	
6th	*AF*	240	

> Copy the diagram and mark the edges as they are chosen.

The next shortest edge is BC, but this forms a cycle with BG and GC, so:

| 7th | *ED* | 280 |

There are eight vertices, and seven edges have been chosen, spanning the tree. The total length to tarmac is 1490 m.

Disadvantages of Kruskal's algorithm

Although Kruskal's algorithm is a simple process and easy to apply to a diagram, it is less easy to apply to a network stored as a matrix. Therefore, the algorithm is not very well suited to computerisation. The problem lies in the need to check at each stage whether the edge about to be chosen forms a cycle with those already chosen. Fortunately, there is a second algorithm, **Prim's algorithm**, which overcomes this problem.

Prim's algorithm

In Prim's algorithm, you only consider the edges joining a vertex that is already connected to a vertex which is yet to be connected. The effect is that your connector 'grows' as a single tree, unlike Kruskal's algorithm, where at an intermediate stage you may have several unconnected parts to your connector.

At each stage the vertices are in two sets, the connected vertices and the unconnected vertices.

Step 1   Choose any vertex to be the first in the connected set.

Step 2   Choose the edge of minimum weight joining a connected vertex to an unconnected vertex. Add this edge to the spanning tree and the vertex to the connected set.

Step 3   If any unconnected vertices remain, go to Step 2.

D1

As with Kruskal's algorithm, when there is a choice between two equal edges at any stage, choose at random.

Prim's algorithm automatically avoids the problem of forming cycles, because it only considers edges to unconnected vertices, which cannot possibly form a cycle.

Example 2

List the order in which Prim's algorithm chooses edges in the country park example, taking the car park as the first vertex.

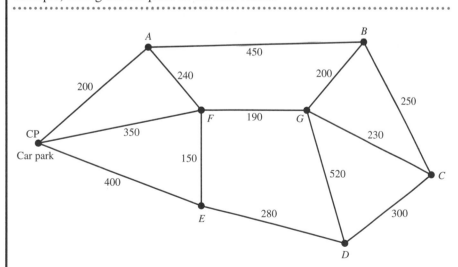

With the car park (*CP*) as the first connected vertex, you choose between the edges *CP–A*, *CP–F* and *CP–E*. The minimum weight is *CP–A* = 200, so you add this edge to your spanning tree, and *A* is now a connected vertex.

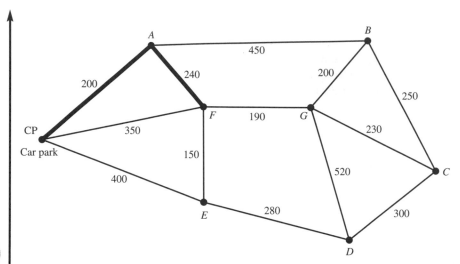

Now choose between the edges *CP–F*, *CP–E*, *A–F* and *A–B*. The minimum weight is *A–F* = 240. So, you add this to your spanning tree, and *F* is now a connected vertex.

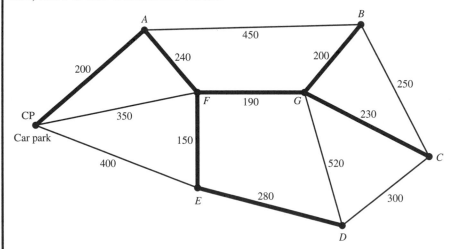

Now choose between the edges connecting to *CP*, *A* or *F*. The minimum weight is *FE* = 150.

Continuing in this way, you go on to choose *FG* = 190, then *BG* = 200, *GC* = 230 and finally *ED* = 280.

The total length is 1490 m, as before.

There are now no vertices left unchosen, so the spanning tree is complete.

Applying Prim's algorithm to a table

The advantage of Prim's algorithm is that it can be applied to a network which has been stored as a matrix. Therefore, the algorithm can be more readily computerised.

The table represents the country park example.

	Car park	A	B	C	D	E	F	G
Car park	–	200	–	–	–	400	350	–
A	200	–	450	–	–	–	240	–
B	–	450	–	250	–	–	–	200
C	–	–	250	–	300	–	–	230
D	–	–	–	300	–	280	–	520
E	400	–	–	–	280	–	150	–
F	350	240	–	–	–	150	–	190
G	–	–	200	230	520	–	190	–

Each stage in Prim's algorithm involves transferring a vertex from the set of unconnected vertices to the set of connected vertices. This is shown on the table by the following devices.

✦   Cross out the row for that vertex (remove it from the unconnected set)

✦   Circle the heading of the column for that vertex (add it to the connected set).

The columns should be labelled 1, 2, 3, … to show the order in which the vertices were chosen.

The solution is demonstrated here, choosing the car park as the first connected vertex. Cross out the car park row, circle the column for the car park and label it **1**, as shown below.

	1							
	(Car park)	A	B	C	D	E	F	G
~~Car park~~	~~–~~	~~200~~	~~–~~	~~–~~	~~–~~	~~400~~	~~350~~	~~–~~
A	200	–	450	–	–	–	240	–
B	–	450	–	250	–	–	–	200
C	–	–	250	–	300	–	–	230
D	–	–	–	300	–	280	–	520
E	400	–	–	–	280	–	150	–
F	350	240	–	–	–	150	–	190
G	–	–	200	230	520	–	190	–

D1

The edge from the car park with the minimum weight is to A, so choose this edge. Circle the 200, cross through the A row, and circle and label the A column, giving:

	1	2						
	Car park	A	B	C	D	E	F	G
~~Car park~~	–	200	–	–	–	400	350	–
~~A~~	(200)	–	450	–	–	–	240	–
B	–	450	–	250	–	–	–	200
C	–	–	250	–	300	–	–	230
D	–	–	–	300	–	280	–	520
E	400	–	–	–	280	–	150	–
F	350	240	–	–	–	150	–	190
G	–	–	200	230	520	–	190	–

> Label the column for vertex A with a **2**.

The available edges appear in the columns labelled **1** and **2**. The minimum weight is $AF = 240$. AF is therefore the next chosen edge. Circle the 240, cross through the F row, and circle and label the F column, giving:

	1	2					3	
	Car park	A	B	C	D	E	F	G
~~Car park~~	–	200	–	–	–	400	350	–
~~A~~	(200)	–	450	–	–	–	240	–
B	–	450	–	250	–	–	–	200
C	–	–	250	–	300	–	–	230
D	–	–	–	300	–	280	–	520
E	400	–	–	–	280	–	150	–
~~F~~	350	(240)	–	–	–	150	–	190
G	–	–	200	230	520	–	190	–

> Label the column for vertex F with a **3**.

The available edges appear in the columns labelled **1**, **2** and **3**. The minimum weight is $EF = 150$. Circle the 150, cross through the E row, and circle and label the E column, giving:

	1	2				4	3	
	Car park	A	B	C	D	E	F	G
~~Car park~~	–	200	–	–	–	400	350	–
~~A~~	(200)	–	450	–	–	–	240	–
B	–	450	–	250	–	–	–	200
C	–	–	250	–	300	–	–	230
D	–	–	–	300	–	280	–	520
~~E~~	400	–	–	–	280	–	(150)	–
~~F~~	350	(240)	–	–	–	150	–	190
G	–	–	200	230	520	–	190	–

> Continue labelling the columns in order as the vertices are chosen.

The available edges appear in the columns labelled **1**, **2**, **3** and **4**. The minimum weight is $FG = 190$. Circle the 190, cross through the G row, and circle and label the G column, giving:

	1	2				4	3	5
	(Car park)	(A)	B	C	D	(E)	(F)	(G)
~~Car park~~	–	200	–	–	–	400	350	–
~~A~~	(200)	–	450	–	–	–	240	–
B	–	450	–	250	–	–	–	200
C	–	–	250	–	300	–	–	230
D	–	–	–	300	–	280	–	520
~~E~~	400	–	–	–	280	–	(150)	–
~~F~~	350	(240)	–	–	–	150	–	190
~~G~~	–	–	200	230	520	–	(190)	–

The available edges appear in the columns labelled **1**, **2**, **3**, **4** and **5**. The minimum weight is $BG = 200$. Circle the 200, cross through the B row, and circle and label the B column, giving:

	1	2	6			4	3	5
	(Car park)	(A)	(B)	C	D	(E)	(F)	(G)
~~Car park~~	–	200	–	–	–	400	350	–
~~A~~	(200)	–	450	–	–	–	240	–
~~B~~	–	450	–	250	–	–	–	(200)
C	–	–	250	–	300	–	–	230
D	–	–	–	300	–	280	–	520
~~E~~	400	–	–	–	280	–	(150)	–
~~F~~	350	(240)	–	–	–	150	–	190
~~G~~	–	–	200	230	520	–	(190)	–

The available edges appear in the columns labelled **1**, **2**, **3**, **4**, **5** and **6**. The minimum weight is $CG = 230$. Circle the 230, cross through the C row and circle the C column, giving:

	1	2	6	7		4	3	5
	(Car park)	(A)	(B)	(C)	D	(E)	(F)	(G)
~~Car park~~	–	200	–	–	–	400	350	–
~~A~~	(200)	–	450	–	–	–	240	–
~~B~~	–	450	–	250	–	–	–	(200)
~~C~~	–	–	250	–	300	–	–	(230)
D	–	–	–	300	–	280	–	520
~~E~~	400	–	–	–	280	–	(150)	–
~~F~~	350	(240)	–	–	–	150	–	190
~~G~~	–	–	200	230	520	–	(190)	–

D1

The available edges appear in the columns labelled **1**, **2**, **3**, **4**, **5**, **6** and **7**. The minimum weight is $DE = 280$. Circle the 280, cross through the D row, and circle and label the D column, giving:

	1	2	6	7	8	4	3	5
	(Car park)	(A)	(B)	(C)	(D)	(E)	(F)	(G)
~~Car park~~	–	~~200~~	–	–	–	~~400~~	~~350~~	–
~~A~~	(200)	–	~~450~~	–	–	–	~~240~~	–
~~B~~	–	~~450~~	–	~~250~~	–	–	–	(200)
~~C~~	–	–	~~250~~	–	~~300~~	–	–	(230)
~~D~~	–	–	–	~~300~~	–	(280)	–	~~520~~
~~E~~	~~400~~	–	–	–	~~280~~	–	(150)	–
~~F~~	~~350~~	(240)	–	–	–	~~150~~	–	~~190~~
~~G~~	–	–	~~200~~	~~230~~	~~520~~	–	(190)	–

Now all the columns are numbered and all the rows are crossed through. The total length is 1490 m as before.

All the vertices are now connected, so the process is complete. The circles in the table show the edges forming the minimum spanning tree, and the numbers at the top show the order in which they were chosen.

Prim's algorithm for a matrix can be summarised as follows.

Step 1   Select the first vertex.

Step 2   Cross out the row for the chosen vertex. Circle and number the column heading for the chosen vertex.

Step 3   Find the minimum, uncrossed-out weight in the columns beneath the circled vertices. Circle this value and note the row in which it lies. The vertex for this row is chosen next.

Step 4   Repeat steps 2 and 3 until all vertices have been chosen.

As usual, when you are faced with a choice between two equal weights, choose at random.

> You might like to check that the process chose the same vertices and edges in the same order as when Prim's algorithm was applied to the network diagram (page 13).

> Look in all the circled columns, not just the one most recently circled.

Exercise 2A

1   Use Kruskal's algorithm to find the minimum connector for each of the two networks shown. List the order in which you choose the edges, and find the total weight of each connector.

a)

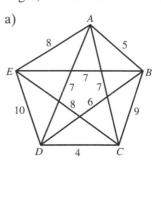

b)

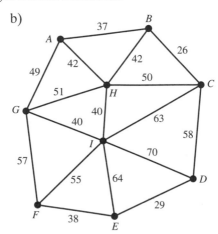

2 Use Kruskal's algorithm to find the two possible minimum spanning trees for the network shown. List the order in which you choose the edges, and find the total weight of the connector.

3 A farmer has five animal shelters on his land and wishes to connect them all to the water supply available at the farmhouse. The table shows the distances (in metres) between the farmhouse, F, and the shelters A, B, C, D and E (some direct connections are not possible).

a) Draw a network diagram to match the table.

b) Use Kruskal's algorithm to find which connections the farmer should make to achieve the water supply as efficiently as possible. List the order in which you choose the connections and find the total length of water piping the farmer would need.

	A	B	C	D	E	F
A	–	70	100	120	80	50
B	70	–	–	–	70	–
C	100	–	–	60	–	–
D	120	–	60	–	–	80
E	80	70	–	–	–	–
F	50	–	–	80	–	–

D1

4 Use Prim's algorithm to find the minimum spanning tree for each of the networks shown. In each case, use A as the starting vertex and list the order in which the vertices are connected. Find the total weight of the spanning tree.

a)

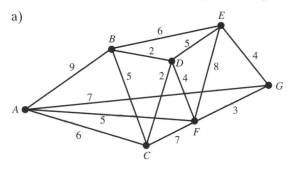

b)

5 Use Prim's algorithm on the table in Question **3**, with the farmhouse as the starting vertex. List the order in which the vertices are connected.

6 The table (right) shows the distance by direct rail link between eight towns. During a period of rationalisation, it is decided to close some of the links, leaving just enough connections so that it is possible to travel from any town to any other by rail. Use Prim's algorithm to decide which links must be kept so that the amount of track is a minimum. Use town A as the starting vertex and record the order in which you make the links.

	A	B	C	D	E	F	G	H
A	–	56	20	-	–	–	–	70
B	56	–	–	15	65	–	75	88
C	20	–	–	87	95	–	120	30
D	–	15	87	–	60	–	25	112
E	–	65	95	60	–	30	40	70
F	–	–	–	–	30	–	45	–
G	–	75	120	25	40	45	–	115
H	70	88	30	112	70	–	115	–

7 The diagram on the right shows the roads connecting five villages, together with distances in kilometres.

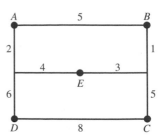

a) Construct a table showing the distances by road between the five villages (for example, the distance from A to E is 6 km).

b) Use Prim's algorithm, starting from vertex A, to find the minimum connector for your table.

As an environmental measure, the local council decides to make some stretches of road 'pedestrians, horses and cycles only', leaving just the minimum unrestricted road needed to enable cars to travel between the five villages.

c) Explain why the result you obtained in part b) does not represent the number of kilometres of road which must remain open to cars.

d) Use the original diagram to find by inspection the roads which should be kept open to cars.

8 A gardener has a patio area paved with square 50 cm slabs. She intends to install a number of small lights, for which she must have a power supply. She plans to hide the cable in the cracks between the paving slabs. Cable can go from light to light, but there are to be no joins in the cable at other points.

The diagram shows the power source at O and the lights at A, B, C, D, E and F.

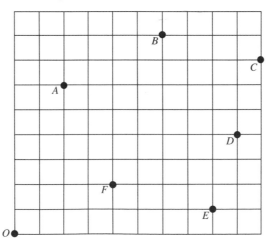

a) Ignoring the size of the gaps between the slabs, complete the table for the length of cable run (in metres) between the various lights.

	O	A	B	C	D	E	F
O	–	4					
A	4	–					
B			–	2.5			
C			2.5	–			
D					–		
E						–	
F							–

b) Use Prim's algorithm, starting from O, to find the best layout for the cable. Show all your working and state how much cable will be needed.

c) Investigate whether it is possible to improve on this solution if the cables may be joined together at other places.

9 The table shows the cost, in pounds per thousand words, of translating between a number of languages.

	English	French	German	Italian	Portuguese	Spanish
English	–	25	30	27	38	22
French	25	–	35	22	36	28
German	30	35	–	35	40	32
Italian	27	22	35	–	26	20
Portuguese	38	36	40	26	–	23
Spanish	22	28	32	20	23	–

a) Use Prim's algorithm to find the minimum spanning tree for this network.

b) From your solution to part (a), state the cheapest sequence of translations to make a document of 10 000 words, originally in German, available in all the languages, and calculate the cost of the operation.

c) Calculate the cost of translating the document directly from German into each of the other languages. Suggest reasons why, despite the extra cost, this might be a preferable course of action.

10 The diagram shows an estuary with five islands. The islands and the mainland are connected by ferries or toll bridges, and the figures shown give the prices, in pounds, of a three-month pass for each crossing.

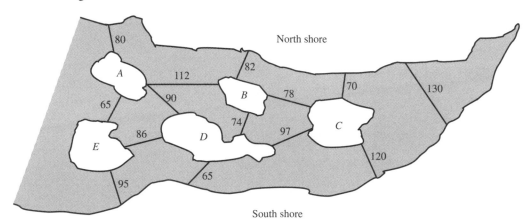

A delivery firm needs to deliver to all areas. Draw a network to model the situation and use Kruskal's algorithm to decide which passes the firm should purchase to give them the desired access as cheaply as possible.

11 The diagram shows the weight limit
(in tonnes) imposed on heavy goods
vehicles on roads between nine towns.
The local council proposes to prohibit
lorries on as many roads as possible in
the area, while maintaining lorry access
to all towns.

By a suitable modification to Prim's
algorithm, choose the roads which the
council should keep open to lorry traffic,
and state the heaviest vehicle which would
have access to all the towns under your
proposal.

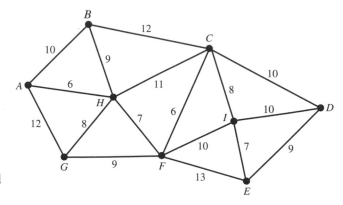

D1 **12** An alternative version of Kruskal's algorithm is as follows.

Starting with the original network of n vertices, and choosing at
random if, at any stage, there is a choice of edges:

    Step 1   Delete the edge with the greatest weight consistent
             with the network remaining connected.

    Step 2   If more than $(n-1)$ edges remain, go to Step 1.

The diagram shows the network from Question **1** b).

Apply this revised version of Kruskal's algorithm to this network,
listing the order in which you delete the edges.

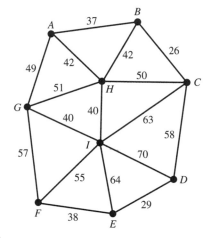

- -

Summary

You should know how to ...

1 Recognise a **cycle**, which is a closed path. That is, a continuous
sequence of edges, with no repeat edges or vertices, leading back to
the starting vertex.

2 Recognise a **tree**, which is a connected graph with no cycles.

3 Recognise a **connector** or **spanning tree** as a tree which includes
every vertex of the graph.
A spanning tree for a graph with n vertices will have $(n-1)$ edges.

4 Identify a **minimum connector** or **minimum spanning tree** for a
network. It is the connector with the least total weight.

5 Apply **Kruskal's algorithm** to a network diagram to find the minimum connector. The algorithm is

Step 1 Choose the edge with the lowest weight.

Step 2 Choose from the remaining edges the one with the lowest weight, provided that it does not form a cycle with those edges already chosen.

Step 3 If any vertices remain unconnected, go to Step 2.

If at any stage there is a choice of edges, choose at random.

6 Apply **Prim's algorithm** to a network diagram to find the minimum connector. The algorithm is

Step 1 Choose any vertex to be the first in the connected set.

Step 2 Choose the edge of the lowest weight joining a connected vertex to an unconnected vertex. Add this edge to the spanning tree and the vertex to the connected set.

Step 3 If any unconnected vertices remain, go to Step 2.

If at any stage there is a choice of edges, choose at random.

7 Apply **Prim's algorithm** to a network expressed as a matrix. The algorithm is

Step 1 Select the first vertex.

Step 2 Cross out the row for the chosen vertex. Circle and number the column heading for the chosen vertex.

Step 3 Find the minimum, uncrossed-out weight in the columns beneath circled vertices. Circle this value and note the row in which it lies. The vertex for this row is chosen next.

Step 4 Repeat steps 2 and 3 until all vertices have been chosen.

If at any stage there is a choice between equal weights, choose at random.

D1

Revision exercise 2

1 The distances, in miles, between six towns are given by this matrix.

	A	B	C	D	E	F
A	–	3	5	10	13	19
B	3	–	4	7	13	23
C	5	4	–	7	10	22
D	10	7	7	–	17	18
E	13	13	10	17	–	9
F	19	23	22	18	9	–

a) Using Prim's algorithm and showing your working at each stage, find a minimum spanning tree for these six towns.

b) State the length of your minimum spanning tree.

c) When information is provided in matrix form, explain why Prim's algorithm, in preference to Kruskal's algorithm, is normally used to find a minimum spanning tree. *(AQA, 1999)*

2 A company, which organises coach holidays, arranges for its passengers to be picked up by feeder coaches at a number of collection points and brought to a central interchange. They are then transferred to the correct coach for their holiday.

The network shows the places where there are collection points and the distances in miles between them.

a) Use an algorithm to find a minimum spanning tree for the network.

b) The company has its central interchange at K and uses a minimum spanning tree for the routes of its feeder coaches.
   i) Suggest a modification to your spanning tree found in part a) which would reduce the distance travelled by some of the feeder coaches.
   ii) Explain, with reference to the network, why the company might prefer K to W as the location for its central interchange. *(NEAB/AQA, 1997)*

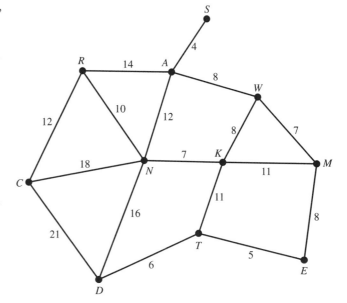

3 The walking times, in minutes, between six points A–F are given in the table.

	A	B	C	D	E	F
A	–	20	30	20	15	10
B	20	–	35	30	25	25
C	30	35	–	30	30	25
D	20	30	30	–	15	25
E	15	25	30	15	–	10
F	10	25	25	25	10	–

a) Use Prim's algorithm, starting at A, to find a minimum connector of the six points.

b) Draw a tree representing the minimum connector which you found in part a).

c) Due to poor weather, only the footpaths in the minimum connector are open. Use your tree from part b) to show that it is still possible to walk from any one of the six points to any other in less than 1 hour.

(AQA, 2004)

4 The table shows the distances, in miles, of any direct routes between the seven towns A, B, C, D, E, F and G.

	A	B	C	D	E	F	G
A	–	5	15	–	–	–	7
B	5	–	8	–	–	9	15
C	15	8	–	7	13	–	6
D	–	–	7	–	5	6	10
E	–	–	13	5	–	5	–
F	–	9	–	6	5	–	6
G	7	15	6	10	–	6	–

D1

a) Use Prim's algorithm to find a minimum connector of the seven towns, and state its length.

b) The local authority decides to charge tolls on some of these roads, but it must still be possible to travel between any two towns on toll-free roads. What is the maximum length of roads on which it can charge tolls?

c) Opposition from local residents forces the authority to keep the road from B to G free of tolls (and to leave enough other toll-free roads so that it is still possible to travel between any two towns on toll-free roads). With this extra restriction, what is the maximum length of roads on which the local authority can charge tolls?

(AQA, 2002)

5 A connected graph G has five vertices. The lengths of the edges are 7, 7, 7, 9, 11, 12, 13, 15 and 16 units, respectively.

a) A graph is said to be *fully connected* if every vertex is joined at least once to every other vertex. Explain why this graph G cannot be fully connected.

b) A connected graph G is to be drawn with five vertices with lengths of edges as given above.
   i) Calculate the least possible length of a minimum spanning tree of graph G.
   ii) Explain why your answer to part b) i) might not apply to graph G.
   iii) Sketch an example of graph G which has a minimum spanning tree of length 34 units.

(AQA, 1999)

D1

6 The table on the right shows the lengths, in miles, of pipelines between six water-pumping stations A–F.

	A	B	C	D	E	F
A	–	23	17	19	18	22
B	23	–	20	24	23	21
C	17	20	–	24	19	23
D	19	24	24	–	20	24
E	18	23	19	20	–	24
F	22	21	23	24	24	–

a) The authorities decide to disconnect some of these pipelines. Use Prim's algorithm, starting at A, to find the minimum total length of pipelines which need to be kept open in order for all the pumping stations to remain linked together.

b) The pipeline from B to C is found to be faulty. So, when choosing the pipelines to keep open, the authorities must not include BC. By how many miles does this increase the total length of pipelines which need to be kept open?

(AQA, 2003)

7 An international organisation has offices A, B, C, D, E, F, G and H. The table shows the cost, in pounds, of transmitting a piece of information from one of the offices to another along all existing direct links.

From \ To	A	B	C	D	E	F	G	H
A	–	12	8	9	–	–	–	–
B	12	–	6	–	10	–	–	–
C	8	6	–	10	–	10	–	–
D	9	–	10	–	–	–	12	–
E	–	10	–	–	–	11	–	20
F	–	–	10	–	11	–	18	18
G	–	–	–	12	–	18	–	14
H	–	–	–	–	20	18	14	–

a) Construct a network using the vertices A to H to represent the information in the table.

b) A piece of information has to be passed from office A to all the other offices, either directly or by being passed on from office to office. Use Kruskal's algorithm to find the minimum total cost of passing the information to all the offices.

c) Office C joins a communications discount scheme in which the cost of passing information from C is halved (but it does not affect the cost of C's incoming information).
   i)   Describe how to adapt the above table to show the revised costs.
   ii) Adapt your answer to part b) in order to calculate the minimum cost of passing a piece of information from office A to all the other offices with these revised costs.

(AQA, 2002)

D1

3 Shortest path

This chapter will show you how to

♦ Recognise when a problem involves finding the shortest route between two vertices of a network
♦ Apply Dijkstra's algorithm to finding a shortest route

3.1 Finding the shortest route

You are probably familiar with route-planning software which enables a motorist to choose the shortest route between two places. The application of this type of package is also important for transport and distribution companies.

Software for finding the shortest route cannot examine every possible route before choosing the best, because for all but the most trivial networks the number of routes is very large. Instead, it makes use of an algorithm which systematically builds the best route.

> There are many situations in which the solution of the problem requires you to find the route with the lowest total weight – the **shortest route** – between two vertices of a network. The weights may be distances but could instead be times, costs, and so forth.

Dijkstra's algorithm

One such algorithm is Dijkstra's (pronounced *Dyke-stra's*) algorithm. It is a labelling algorithm and works by moving through the network from the starting vertex, labelling the vertices with temporary 'best distance found so far' labels. These labels are then made permanent when it is clear that they represent the shortest route to that vertex. At each stage, another vertex receives this permanent label, so that when the algorithm is finished every vertex is labelled with its shortest distance from the start.

> This was developed by Edsger Dijkstra (1930–2002) in 1959.

Before formally stating the algorithm, look at its application to a simple network. The task is to find the shortest route from vertex S to vertex T in the network on the right. (The start and finish, or terminus, vertices are often denoted by S and T.)

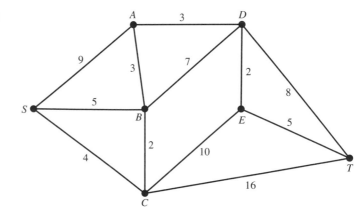

Obviously, the shortest distance from S to S is zero. So, at the first stage you can give S the permanent label 0. You indicate that a label is permanent by drawing a box round it. You then give temporary labels to all vertices which can be reached directly from S. The labels are:

$0 + 9 = 9$ at A
$0 + 5 = 5$ at B
$0 + 4 = 4$ at C

as shown.

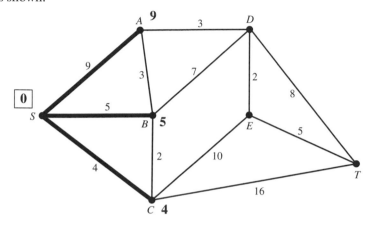

The shortest route to C is clearly 4, since any other route would have to pass through A or B and would be longer. At the second stage, therefore, you give C the permanent label 4.

You then give temporary labels to all vertices which can be reached from C and which are not permanently labelled. If a vertex already has a temporary label, you only replace this if it is an improvement. Consequently, the temporary label for B remains at 5, since its possible replacement $(4 + 2)$ is not an improvement. The labels you add are:

$4 + 16 = 20$ at T
$4 + 10 = 14$ at E

as shown.

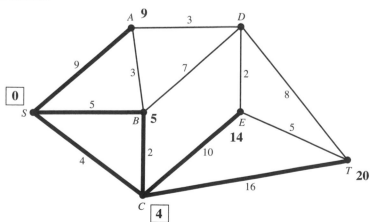

B now has the smallest temporary label. The shortest route to B must therefore be 5, since any other route would have to pass through a vertex with a larger label. At the third stage, therefore, you give B the permanent label 5.

You then give temporary labels to all vertices which can be reached from B and which are not permanently labelled, replacing temporary labels where there is an improvement. The labels you add are:

$5 + 3 = 8$ replacing 9 at A
$5 + 7 = 12$ at D

as shown.

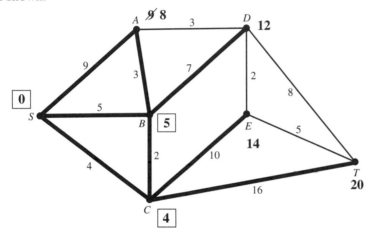

A now has the smallest temporary label. So, at the fourth stage, you give A the permanent label 8.

You then give temporary labels to all vertices which can be reached from A, replacing temporary labels where there is an improvement. The label you add is:

$8 + 3 = 11$ replacing 12 at D

as shown.

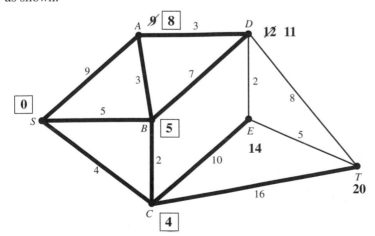

D now has the smallest temporary label. So, at the fifth stage, you give *D* the permanent label 11.

The temporary labels are *D*, 11; *E*, 14 and *T*, 20.

You then give temporary labels to all vertices which can be reached from *D*, replacing temporary labels where there is an improvement. The labels you add are:

11 + 2 = 13 replacing the 14 at *E*
11 + 8 = 19 replacing the 20 at *T*

as shown.

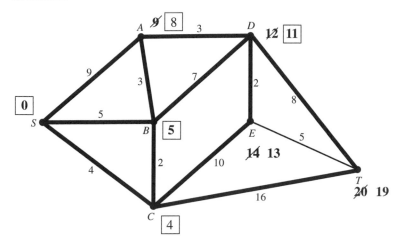

E now has the smallest temporary label. So, at the sixth stage you give *E* the permanent label 13.

E has temporary label, 13, and *T* has temporary label, 20.

You then give temporary labels to all vertices which can be reached from *E*, replacing temporary labels where there is an improvement. The label you add is:

13 + 5 = 18 replacing the 19 at *T*

as shown.

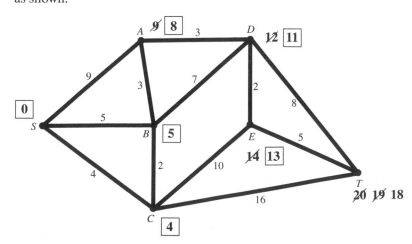

The final stage is to give T the permanent label 18, as shown.

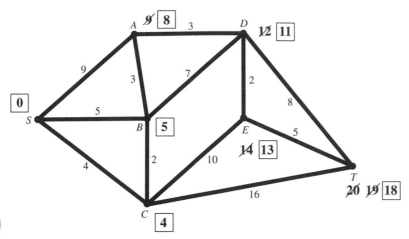

D1

Each permanent label represents the shortest distance from S to that vertex. The shortest route to T is therefore 18, and it just remains to determine what that route is.

You work backwards from T. To be part of the route, the weight of an edge must equal the difference between the labels of the two vertices. Hence, you have:

Label T – Label E = 18 – 13 = 5 = Weight of ET,
  so ET is part of the route.
Label E – Label D = 13 – 11 = 2 = Weight of DE,
  so DE is part of the route.
Label D – Label A = 11 – 8 = 3 = Weight of AD,
  so AD is part of the route.
Label A – Label B = 8 – 5 = 3 = Weight of BA,
  so BA is part of the route.
Label B – Label S = 5 – 0 = 5 = Weight of SB,
  so SB is part of the route.

The shortest route is, therefore, $SBADET$ = 18.

In some textbooks, you will see the labelling recorded in a series of boxes, with spaces for the temporary labels, the permanent label and a number representing the order in which the vertices were chosen, as shown below. However, the labelling system used in this example is completely acceptable for the AQA syllabus.

Stage	Permanent label
Temporary labels	

Formal statement of Dijkstra's algorithm

Step 1   Label the start vertex with permanent label 0.

Step 2   If V is the most recent vertex to receive a permanent label, update the temporary labels of all vertices directly connected to V.

Each of these vertices gets:

   Temporary label = (Label of V + Weight of connecting edge)

unless it already has a temporary label less than or equal to this value.

Step 3   Choose the vertex with the lowest temporary label (choose at random if there is a tie) and permanently label it with this value. If this is the destination vertex then stop. Otherwise, go to Step 2.

You trace back to find the route. If vertex Y is on the route, you find the neighbouring vertex X such that:

Label Y – Label X = Weight of edge XY.

Vertex X is then the previous vertex on the route.

3.2 Directed networks

Some networks involve 'one-way streets' – edges which may only be travelled in one direction. The graph is then called a **directed graph** or a **digraph**. Even if it is possible to travel directly between two vertices in both directions, the weights may be different. An example would be the time to cycle from the bottom to the top of a hill, and from the top to the bottom. In such a case, the network will have two directed edges connecting the vertices.

The adjacency matrix for a digraph and the distance matrix for a directed network will not be symmetrical. For example:

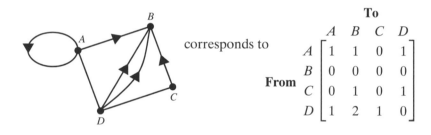

corresponds to

$$\text{From} \begin{array}{c} \\ A \\ B \\ C \\ D \end{array} \begin{array}{c} \overset{\textstyle\textbf{To}}{\begin{array}{cccc} A & B & C & D \end{array}} \\ \left[\begin{array}{cccc} 1 & 1 & 0 & 1 \\ 0 & 0 & 0 & 0 \\ 0 & 1 & 0 & 1 \\ 1 & 2 & 1 & 0 \end{array} \right] \end{array}$$

> It is vital to show all your working in examination questions. The criterion is not whether you can find the shortest route (which is probably quite obvious in such a small-scale problems) but whether you can correctly apply Dijkstra's algorithm.

D1

> The zeros in row B show that there are no routes out of B. The 2 in row D shows that there are 2 routes from D to B.

You can still use Dijkstra's algorithm with directed networks.

Example 1

Find the shortest route between S and T for this network.

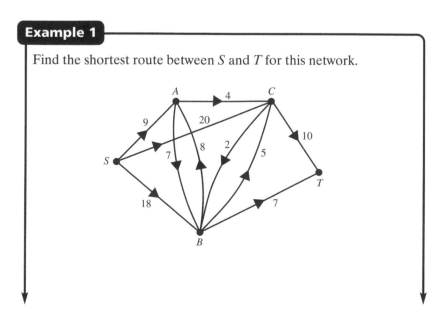

You give S the permanent label 0, then give temporary labels to the adjacent vertices, as shown.

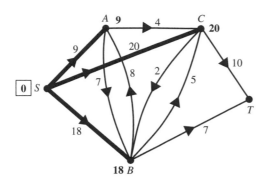

A, B and C can be reached directly from S.

D1

A has the lowest temporary label. So, you make it permanent and update the temporary labels of the adjacent vertices, as shown.

B and C can be reached directly from A.

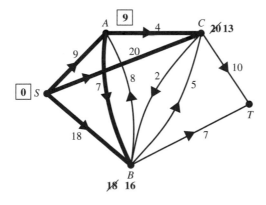

C now has the lowest temporary label. So you make it permanent and update the temporary labels of the adjacent vertices, as shown.

The temporary labels are B, 15 and C, 13.

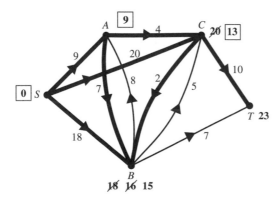

B now has the lowest temporary label. So you make this permanent and update the temporary labels of the adjacent vertices, as shown.

T and *B* can be reached from *C*. The temporary labels are *T*, 22 and *B*, 15.

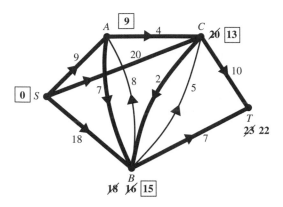

Finally, you make *T* permanent, as shown.

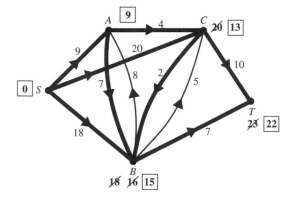

All the vertices now have permanent labels.

You can now state that the shortest route from *S* to *T* is of length 22.

You trace the route back by identifying the edges where the weight equals the difference between the permanent labels of its two vertices:

> Weight *BT* = 7 and (Label *T* – Label *B*) = 7,
> so *BT* is part of the route.
>
> Weight *CB* = 2 and (Label *B* – Label *C*) = 2,
> so *CB* is part of the route.
>
> Weight *AC* = 4 and (Label *C* – Label *A*) = 4,
> so *AC* is part of the route.
>
> Weight *SA* = 9 and (Label *A* – Label *S*) = 9,
> so *SA* is part of the route.

The route is, therefore, *SACBT*.

D1

Limitation of Dijkstra's algorithm

Dijkstra's algorithm fails if any of the weights are negative. To illustrate this, consider the part-network shown on the right.

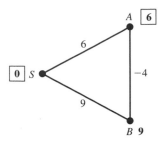

After stage 2 of Dijkstra's algorithm, A has been given a permanent label of 6, as shown. However, the route SBA would give a total weight of $9 + (-4) = 5$, so Dijkstra's algorithm has failed to find the shortest route from S to A.

Negative weights can arise in a variety of ways. For example, the weight of an edge might represent the cost of a lorry travelling that edge. Payment to transport goods along a particular edge might gain a profit, which would appear on that edge as a negative cost.

D1

Exercise 3A

1 Draw the directed graph corresponding to the adjacency matrix shown on the right.

$$\begin{array}{c} & \textbf{To} \\ & \begin{array}{cccc} A & B & C & D \end{array} \\ \textbf{From} \begin{array}{c} A \\ B \\ C \\ D \end{array} & \begin{bmatrix} 1 & 2 & 0 & 0 \\ 0 & 0 & 0 & 1 \\ 0 & 2 & 0 & 1 \\ 0 & 0 & 1 & 0 \end{bmatrix} \end{array}$$

2 Write down the adjacency matrix for the digraph displayed on the right.

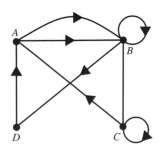

3 Use Dijkstra's algorithm to find the shortest route from S to T on this network. Show all your working. State the route and the distance.

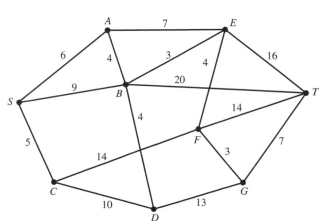

4 Use Dijkstra's algorithm to show that there are two possible shortest routes from S to T on the network below. State the distance and the two possible routes.

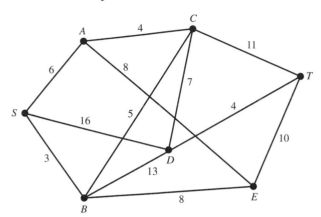

D1

5 The table shows the journey times by bus between a number of towns.

	A	B	C	D	E	F	G
A	–	30	40	18	–	–	–
B	30	–	8	10	40	–	–
C	40	8	–	25	–	15	–
D	18	10	25	–	45	40	–
E	–	40	–	45	–	10	10
F	–	–	15	40	10	–	35
G	–	–	–	–	10	35	–

a) Draw a network diagram to correspond to this matrix.

b) Use Dijkstra's algorithm to find the route from A to G with the lowest total travelling time.

c) Give reasons why this might not actually be the best route to take.

6 Use Dijkstra's algorithm to find the shortest route from A to G, and from G to A, for this directed network.

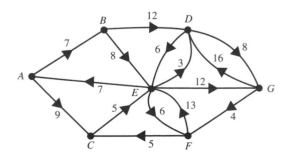

7 A distribution company has three warehouses at towns A, B and C. A customer in town L orders an item, which could be supplied from any of the warehouses. The network shows the distances (in kilometres). Use Dijkstra's algorithm to decide which of the three warehouses should supply the item in order to minimise the delivery distance. State the distance and the route taken. (You are advised to give a little thought to the best strategy for solving this problem.)

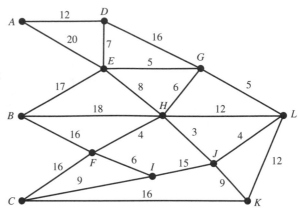

D1 **8** The network shows the cost of air travel (in £) between a number of locations.

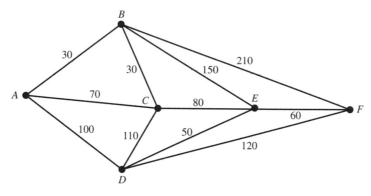

a) Use Dijkstra's algorithm to find the cheapest route from A to F.

b) A new airport duty is introduced so that every intermediate stop is charged at £20. Apply Dijkstra again, adjusting your temporary labels to incorporate this extra cost. What is now the cheapest route?

9 The network shows the weight restrictions (in tonnes) on heavy goods vehicles on a number of roads.

Adapting Dijkstra's algorithm by labelling vertices with the heaviest vehicle which can legally travel to that point, find the heaviest vehicle which can make the journey from A to H, and state the route it should take.

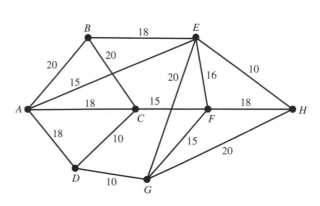

Summary

You should know how to ...

1 Apply **Dijkstra's algorithm** to find the shortest distance between two vertices of a network. The algorithm is:

Step 1 Label the start vertex with permanent label 0.

Step 2 If V is the most recent vertex to receive a temporary label, update the temporary labels of all vertices directly connected to V. Each of these vertices gets:

Temporary label = (label of V + weight of connecting edge) unless it already has a temporary label less than or equal to this value.

Step 3 Choose the vertex with the lowest temporary label (choose at random if there is a tie) and permanently label it with this value. If this is the destination vertex then stop. Otherwise go to Step 2.

2 Trace back to find the shortest route.
If vertex Y is on the route, find the neighbouring vertex X such that:

Label Y − Label X = Weight of edge XY

Vertex X is then the previous vertex on the route.

D1

Revision exercise 3

1 The network shows the distances, in kilometres, of various routes between points S, T, U, V, W, X, Y and Z.

Use Dijkstra's algorithm to find the shortest path from S to Z.   (*AQA, 2001*)

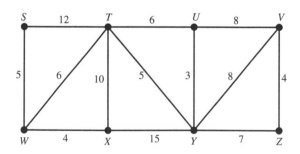

2 The network on the right represents seven campsites P–V and the footpaths between them. The numbers on the edges are the lengths of those footpaths in miles.

a) Use Dijkstra's algorithm to find the shortest route from P to V on footpaths. Show your working clearly at each vertex.

b) Some walkers on a camping holiday spend each night at a campsite and can only walk up to 10 miles a day. What is the minimum number of days which they need to get from P to V? Give a reason for your answer.   (*AQA, 2004*)

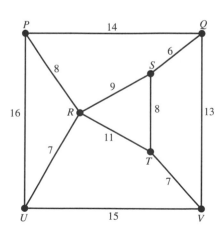

3 The following network shows the time, in minutes, of train journeys between seven stations.

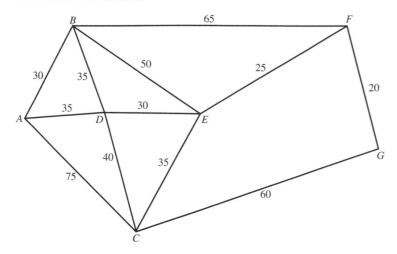

a) Given that there is no time delay in passing through a station, use Dijkstra's algorithm to find the shortest time to travel from A to G.

b) Find the shortest time to travel from A to G, if, in reality, each time the train passes through a station, excluding A and G, an extra 10 minutes is added to the journey time.

(AQA, 1999)

4 The network shows nine towns A to I and the various pairs of them which are connected by bus routes, all provided by the Bunson Bus Company. The number on each edge is the fare, in pence, for using that bus route.

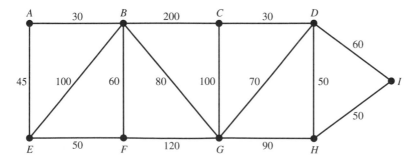

a) Use Dijkstra's algorithm to find the way of getting from A to I by bus at the minimum cost and the routes which must be used.

b) A competing bus company introduces an alternative bus route from G to H and it would like to encourage travellers from A to I to use this service. Calculate the greatest amount it can charge for the new service so that the cheapest way of getting from A to I includes the new service from G to H.

c) As an economy measure the Bunson Bus Company wants to remove many of the bus routes. They want to do this so that it is still possible to travel between any two of the towns by their buses, and they want the total cost of the fares on the remaining routes to be as low as possible.

   i) Use an appropriate algorithm to decide which routes should remain.

   ii) Calculate how much it would then cost to get from A to I.   (AQA, 2001)

5 The network shows the roads around the town of Kester (K) and the times, in minutes, needed to travel by car along those roads.

   a) A motorist wishes to travel from A to C along these roads in the minimum possible time. Use Dijkstra's algorithm on the diagram to find the route the motorist should use and the time that the journey will take. Show all your workings clearly.

   b) The four sections of ring-road AB, BC, CD and DA each require the same amount of time, and next year there will be improvements to the ring-road in order to reduce this time from 30 minutes to m minutes. This will enable the motorist to reduce the minimum time for a journey from A to C by 2 minutes. Find the value of m and state his new route.

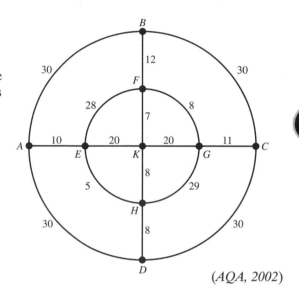

(AQA, 2002)

D1

6 Three boys, John, Lee and Safraz, are to take part in a running race. They are each starting from a different point but they all must finish at the same point N.

John starts from point A, Lee from point B and Safraz from point C.

The diagram shows the network of streets that they may run along. The numbers on the arcs represent the time, in seconds, taken to run along each street.

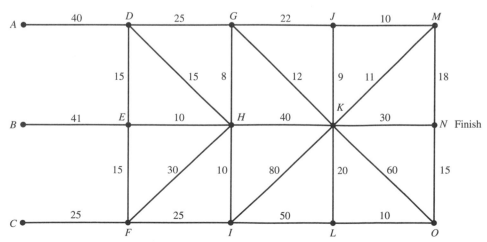

a) Working backwards from N, or otherwise, use Dijkstra's algorithm on the diagram to find the time taken for each of the three boys to complete the course. Show all your working at each vertex.

b) Write down the route that each boy should take.   (*AQA, 2002*)

7 a) The network illustrated has eight vertices:
$A_0, A_1, A_2, A_3, A_{12}, A_{13}, A_{23}$ and A_{123}.

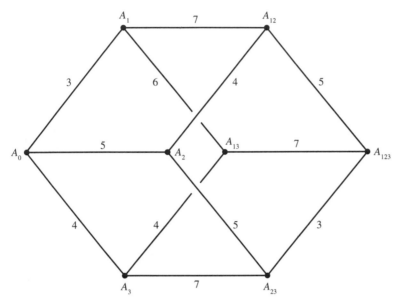

Use Dijkstra's algorithm on the diagram to find the shortest distance from A_0 to A_{123}. Show all your temporary workings clearly.

b) A factory has to produce three items: 1, 2 and 3. The order in which they are produced affects their production costs, as illustrated in the following chart.

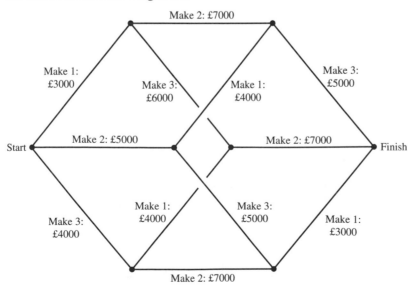

So, for example, if the items are made in the order 1, 3, 2, then the total production cost is £3000 + £6000 + £7000 = £16 000.

Use your answer to part a) to find the minimum total production cost of the three items, and state the order in which the three items should be produced in order to achieve this minimum.

(AQA, 2002)

8 The network below shows the roads from the airport, A, to the Houses of Parliament, H. An overseas president is arriving at the airport and is to be driven with an escort to the Houses of Parliament. The police have identified all security risks, and the number on each arc of the network shows the number of security risks on that road.

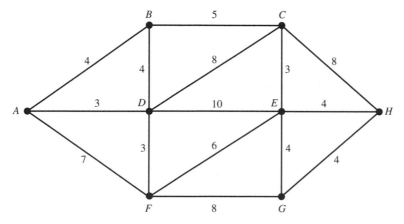

a) Apply Dijkstra's algorithm to find the minimum possible total number of security risks on a route from A to H. Find the **two** routes which have that minimum number of security risks.

b) Just before the president arrives, a bomb scare closes the road EH. Find the route from A to H which should now be used in order to minimise the number of security risks.

(AQA, 2002)

D1

9 A railway company is considering opening some new lines between seven towns *A–G*. The possible lines and the cost of setting them up (in millions of pounds) are shown in the following network.

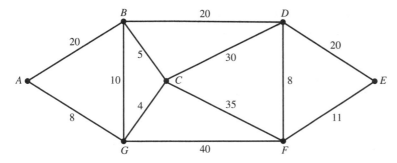

a) Use Dijkstra's algorithm to find the minimum cost of opening lines from *A* to *E*. Show all your workings at each vertex.

b) From your working in part a), write down the minimum cost of opening the following lines:
   i)   from *A* to *B*
   ii) from *A* to *F*

c) Use Kruskal's algorithm to find the minimum cost of opening lines so that it is possible to travel between any two of the towns by rail, and state the lines which should be opened in order to achieve this minimum cost.

(AQA, 2003)

4 Route inspection

This chapter will show you how to

◆ Define the terms traversable (Eulerian) and semitraversable (semi-Eulerian)
◆ Decide whether a given graph is traversable by considering the degree of its vertices
◆ Relate the traversability of a graph to the route inspection problem for a network
◆ Decide for a given network which edges must be repeated to give the optimal solution to the route inspection problem

4.1 The route inspection problem

D1

This is sometimes referred to as the 'Chinese postman problem'. It is the problem, rather than the postman, which is Chinese, having originally been discussed by the Chinese mathematician Mei-ko Kwan in 1962. It can be stated as:

> For a given network, find the shortest route which travels at least once along every edge of the network and returns to the starting point.

Ideally, you want to travel just once along each edge. This is called **traversing** the network. Some networks can be traversed and some cannot. To enable you to decide, you need to learn a little more graph theory.

> Although originally described in terms of a postal round, the problem arises in many contexts. Examples are the planning of routes for gritting lorries on winter roads and the inspection of a network of pipelines.

Traversability and the degree of the vertices

In Chapter 1, page 6, you met the **degree** of a vertex. This is the number of edges which are attached to the vertex. The three graphs below have been labelled with the degree of each vertex.

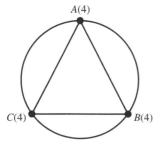

Graph 1

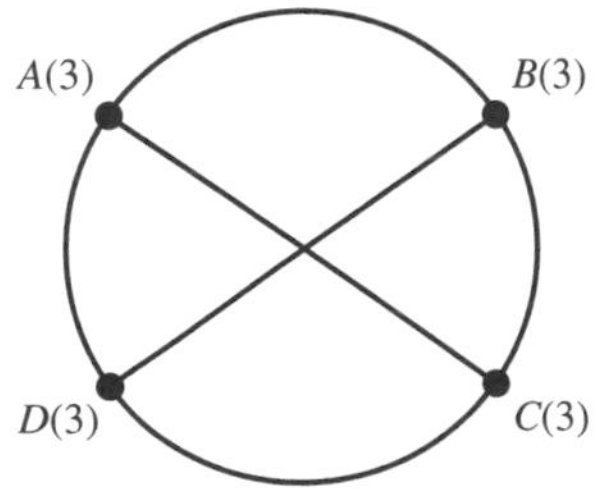

Graph 2

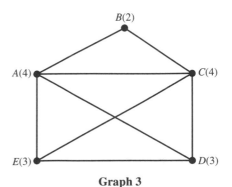

Graph 3

Try to draw each of these graphs without taking your pencil from the paper, ideally starting and finishing in the same place

You should find that Graph 1 can be traversed while Graph 2 cannot. Graph 3 can be drawn, but only by starting at *D* and finishing at *E* (or vice versa). Graph 1 is **traversable** or **Eulerian** (pronounced *oil-eerie-ann*). A route traversing such a graph is called an **Eulerian trail**.

Graph 3 is said to be **semitraversable (semi-Eulerian)**. This means that there is a route which travels once only along each edge, but the start and finish vertices are different.

You can decide about the traversability of a graph by examining the degree of the vertices.

Every time the route passes through a vertex, there must be an 'incoming' edge and an 'outgoing' edge. This means that, apart from the start/finish points, every vertex must have an **even degree**. That is, it must be an **even vertex**.

If the **start vertex** and the **finish vertex** are different, they must each have an even number of edges **plus** the starting **or** finishing edge. So, they must **both** be **odd vertices**.

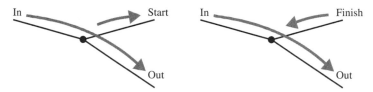

If the **start** and **finish** are at the same vertex, it will have an even number of edges **plus** the starting edge **and** the finishing edge. So, it will be an **even vertex**.

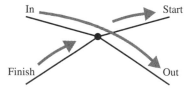

Hence, the degree of the vertices enables you to decide whether a given graph can be traversed, as follows.

> ✦ A graph where **all** the vertices are **even** is **traversable (Eulerian)**.
> ✦ A graph with **two odd vertices** is **semitraversable (semi-Eulerian)**. The two odd vertices are the start and finish points.
> ✦ A graph with more than two odd vertices cannot be traversed.

> Eulerian graphs are named after Leonhard Euler (1708–1783), a Swiss mathematician, among whose many important contributions to mathematics was the study of traversable graphs.

> In an Eulerian graph, each edge is used exactly once and the start and finish are at the same vertex.

> In a semi-Eulerian graph, each edge is used exactly once, but the start and finish are at different odd vertices.

D1

The handshaking theorem

This concerns the sum of the degrees of all the vertices in the graph.

In the same way that one handshake involves two people, so one edge is attached to a vertex at either end. This means that each edge contributes two towards the total of degrees. Hence, you have:

Sum of degrees = 2 × Number of edges

It follows from this that the sum of the degrees is an even number. If you add all the degrees of the vertices and obtain an even total, the number of odd values in the list must be even. Hence, you have:

The number of odd vertices in any graph is even.

Traversability and the route inspection problem

Returning to the route inspection problem, there are two cases to consider:

✦ When all vertices in the network are even, the problem is solved. This is because the network is traversable and the distance that must be travelled is just the sum of the weights of all the edges.

✦ When there are vertices of odd degree, it will be necessary to repeat some edges in order to return to the starting point. In this case, the route inspection problem becomes:

> Which edges must be repeated in order to complete the route as efficiently as possible?

Repeating an edge is equivalent to adding an extra edge to the network diagram. To understand this, consider the following examples.

D1

Example 1

Find a route passing at least once along each edge of this network, starting and finishing at A. Which edge(s) must be repeated in order to achieve this in the minimum distance?

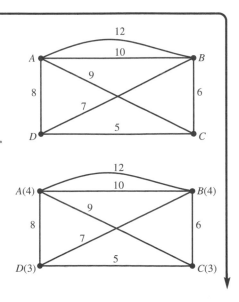

By labelling the vertices with their degrees, you can see that this network is semitraversable. The odd vertices are C and D.

If you try to find a route – for example, $ABCDACDBA$ – you find that it must travel twice between C and D.

This can be illustrated by adding an extra edge, *CD*, to the diagram. The resulting network is then traversable.

Any route traversing this modified network is a possible solution to the problem. The distance to be travelled is

(Sum of original edge weights) + (Additional edge weight)
= 57 + 5 = 62

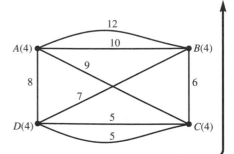

Example 2

Repeat Example 1 for this network, in which the weight of the edge *CD* is 15.

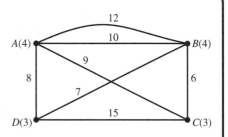

....

You still need to travel twice between the odd vertices *C* and *D*. But, in this case, it is better to make the extra journey via *B*, since this will add (7 + 6) = 13 to the total, rather than the 15 for *CD*.

This can be illustrated by adding extra edges *BC* and *BD* to the diagram. The resulting network is then traversable.

Any route traversing this modified network is a possible solution to the problem. The distance to be travelled is:

(Sum of original edge weights) + (Additional edge weights)
= 67 + (7 + 6) = 80

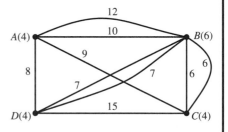

The route is *A B C D B C A B D A*
(12 + 6 + 15 + 7 + 6 + 9 + 10 + 7 + 8 = 80).

General strategy

From Examples 1 and 2, you can see that for a network with one pair of odd vertices, the network is made traversable by adding edges to the diagram corresponding to the shortest route between the odd vertices. (In complex cases, this would involve using Dijkstra's algorithm.) Hence, the total distance to be travelled is then:

(Sum of original edge weights) + (Additional edge weights)

If there are more than two odd vertices (and remember there will always be an even number of them), you can pair them together in various ways. Look for the best way to pair them, so that the extra edges added have the lowest possible total weight.

D1

Example 3

Solve the route inspection problem for this network.

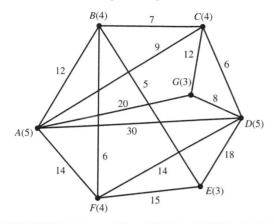

The odd vertices are A, D, E and G. These can be paired together in three ways:

1 AD Shortest route $A–C–D = 15$
     paired with
   EG Shortest route $E–B–C–G = 24$
     Total of extra edges = 15 + 24 = 39

2 AE Shortest route $A–B–E = 17$
     paired with
   DG Shortest route $D–G = 8$
     Total of extra edges = 17 + 8 = 25

3 AG Shortest route $A–G = 20$
     paired with
   DE Shortest route $D–E = 18$
     Total of extra edges = 20 + 18 = 38

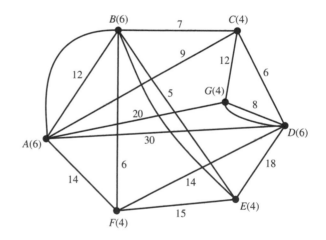

The best pairing is, therefore, AE and DG.
The extra edges are AB, BE and DG, whose total weight is 12 + 5 + 8 = 25.

Total journey = (Sum of original edges) + 25
        = 176 + 25 = 201

A possible route, starting from A, is:

  $ABCDEFABEBFDGCAGDA$

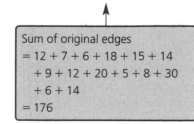

Sum of original edges
= 12 + 7 + 6 + 18 + 15 + 14
  + 9 + 12 + 20 + 5 + 8 + 30
  + 6 + 14
= 176

D1

The route inspection algorithm

The algorithm can now formally be stated as:

    Step 1   Identify the odd vertices.

    Step 2   List all possible pairings of the odd vertices.

    Step 3   For each pairing, find the shortest route between paired vertices. Hence, identify the edges which must be repeated and find the sum of their weights.

    Step 4   Choose the pairing with the smallest sum. By repeating these edges (equivalent to adding extra edges to the network), the network becomes traversable.
            The distance involved is:

         (Sum of original weights) + (Sum of extra edge weights)

D1

The main difficulty with the route inspection problem is that the number of possible pairings increases very rapidly as the number of odd vertices increases. Hence, you have:

2 odd vertices	1 pairing
4 odd vertices	3 pairings
6 odd vertices	15 pairings
8 odd vertices	105 pairings
$\vdots$	$\vdots$
n odd vertices	$(n-1) \times (n-3) \times (n-5) \times \ldots \times 3 \times 1$ pairings

> In the AQA examination, you will not be required to deal with more than four odd vertices. You may, however, be asked how many pairings would need to be considered if the problem had six or more odd vertices.

Variations on the problem

In practical situations, the exact nature of the network to be traversed will vary with the requirements of the problem.

Example 4

The diagram shows a number of streets, total length 2500 m, to be patrolled by a community watch officer. The officer is based at A.

a) Find the best route for the officer to take on foot if each street must be inspected at least once.

b) Find the best route if each side of each street must be inspected, on foot, at least once.

c) Find the best route if the officer is to cycle both ways along each street.

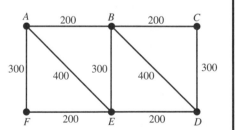

· ·

a) The odd vertices are A and D. The shortest route is $ABD = 600$ m. Adding edges AB and BD to the diagram makes the network traversable.

A possible route is $ABCDEFAEBDBA$. The total distance is $2500 + 600 = 3100$ m.

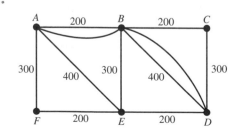

b) Each street must be walked twice. This is equivalent to duplicating each edge on the diagram.

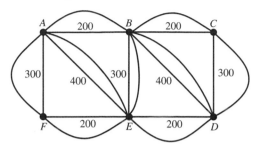

All the resulting vertices are of even degree, so the network is traversable. A possible route is *ABCDEFAEBDEFABCDBEA*. The total distance is $2 \times 2500 = 5000$ m.

D1

c) Each street must be travelled twice but in opposite directions. The distance travelled must, of course, be the same as in part b). However, the route chosen there is unsuitable because it travels *DE*, for example, in the **same** direction on both occasions.

To model this new problem, a directed graph (digraph) must be used, with arrows to show the direction in which each edge may be travelled.

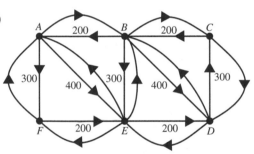

The network is still traversable. (For every incoming edge, there is an outgoing edge.) A possible route is *ABCDEFAEBDCBAFEDBEA*. The total distance of this route is $2 \times 2500 = 5000$ m.

Exercise 4A

1 For the network shown, determine the edges which should be repeated to solve the route inspection problem. State the length of the route and give a possible route starting from vertex *A*.

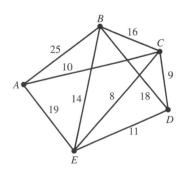

2 Solve the route inspection problem for each of the networks shown. In each case, list the possible pairings of odd nodes, together with the total extra weight these pairings would involve. State the edges to be repeated for the best solution and state the total weight of the resulting route.

a)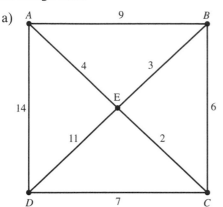

b)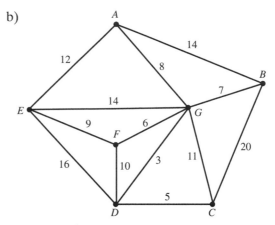

D1

3 The diagram shows the map of the roads on an estate, with distances in metres. All junctions are right angles and all roads are straight apart from the crescent *CD*. A team of workers is to paint a white line along the centre of each road.

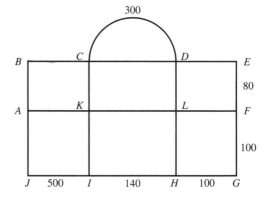

a) Assuming that they will enter and leave the estate at *A* and wish to travel the least possible distance on the estate, determine which sections of road they will need to travel twice and find the distance they will travel.

b) Instead, the team wishes to enter the estate at *A* and leave at *F* to go on to another job. How will this affect your results from part a)?

4 The table shows the direct road links between six towns, with the distances in kilometres. Abigail is to do a sponsored cycle ride, travelling at least once along each of these roads and starting and finishing at her home in town *A*.

	A	*B*	*C*	*D*	*E*	*F*
A	–	15	–	8	20	–
B	15	–	10	–	6	5
C	–	10	–	9	–	14
D	8	–	9	–	9	4
E	20	6	–	9	–	–
F	–	5	14	4	–	–

Draw the network diagram corresponding to this table and hence investigate which roads she should repeat and how long her journey will be.

5 The diagram shows the layout of paths in a small pedestrian shopping precinct. Distances are in metres.

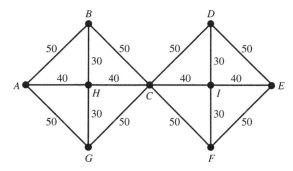

It is necessary to drive a cleaning cart around the precinct, starting and ending at *A*.
Find:
a possible best route to take,
the sections which will be travelled twice, and
the total distance.

D1

6 The diagram shows a logo for a sports goods company, displayed outside a shop. The lengths shown are in centimetres and the logo is symmetrical, with all angles right angles.

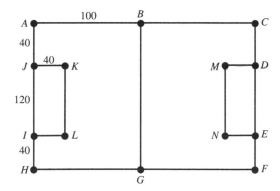

The shop owner wishes to outline it with a single continuous strand of flashing lights, starting and ending at *A*. Calculate the minimum length of cable involved, and state which sections of the logo will have a double run of cable.

Summary

You should know how to ...

1 Identify a traversable (Eulerian) graph, for which it is possible to travel along every edge once and once only, starting and finishing at the same vertex. A graph is traversable if all vertices have an even degree.

2 Identify a semitransversable (semi-Eulerian) graph, for which each edge can be travelled once and once only but the start and finish vertices are different. A graph is semitraversable if two vertices have an odd degree.

3 Solve the route inspection problem using the following algorithm.

D1

> **Step 1** Identify the odd vertices (there is always an even number of odd vertices).
>
> **Step 2** List all possible pairings of the odd vertices.
>
> **Step 3** For each pairing find the shortest route between paired vertices. Hence identify the edges which must be repeated, and find the sum of their weights.
>
> **Step 4** Choose the pairing with the smallest sum. By repeating these edges (equivalent to adding extra edges to the network), the network becomes traversable. The distance involved is:
> (Sum of original weights) + (Sum of extra edge weights)

Revision exercise 4

1 A road gritting service is based at a point A. It is responsible for gritting the network of roads shown in the diagram, where the distances shown are in miles.

a) Explain why it is **not** possible to start from A and, by travelling along each road only once, return to A.

b) In the network there are four odd vertices, B, D, F and G. List the different ways in which these odd vertices can be arranged as two pairs.

c) For **each** pairing you have listed in part b), write down the sum of the shortest distance between the first pair and the shortest distance between the second pair.

d) Hence find an optimal 'Chinese postman' route around the network, starting and finishing at A. State the length of your route.

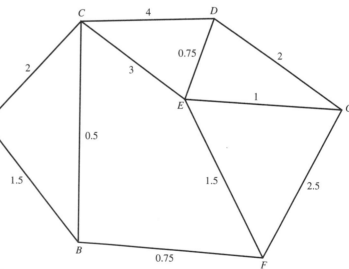

(*AQA, 1998*)

2 A theme park employs a student to patrol the paths and collect litter. The paths that she has to patrol are shown in the diagram, where all the distances are in metres. The path connecting *I* and *W* passes under the bridge which carries the path connecting *C* and *R*.

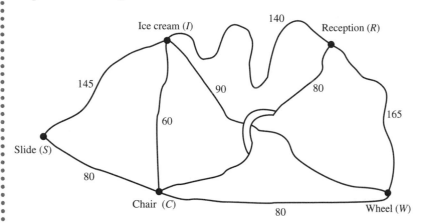

a) i) Find an optimal 'Chinese postman' route that the student should take if she is to start and finish at Reception (*R*).
   ii) State the length of your route.

b) i) A service path is to be constructed. Write down the two places that this path should connect, if the student is to be able to walk along every path without having to walk along any path more than once.
   ii) The distance walked by the student in part b)i) is shorter than that found in part a)ii). Given that the length of the service path is *l* metres, where *l* is an integer, find the greatest possible value of *l*.

(AQA, 1999)

3 A highways department has to inspect its roads for fallen trees.

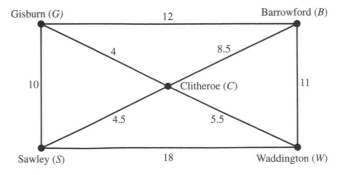

a) The diagram on the right shows the lengths of the roads, in miles, that have to be inspected in one district.
   i) List the three different ways in which the four odd vertices in the diagram can be paired.
   ii) Find the shortest distance that has to be travelled in inspecting all the roads in the district, starting and finishing at the same point.

b) The connected graph of the roads in another district has six odd vertices. Find the number of ways of pairing these odd vertices.

c) For a connected graph with *n* odd vertices, find an expression for the number of ways of pairing these odd vertices.

(AQA, 1999)

D1

4 The network on the right has 16 vertices.

The length of each edge is 1 unit.

a) Find the length of an optimal 'Chinese postman' route, starting and finishing at A.

b) For such a route, state the edges that would have to be used twice.

c) Given that the edges AE and LP are now removed, find the new length of an optimal 'Chinese postman' route, starting and finishing at A. *(AQA, 2003)*

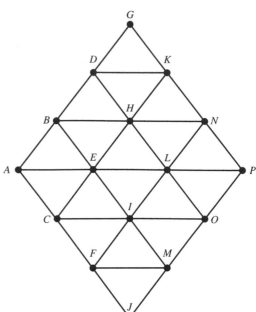

D1

5 The following question refers to the three graphs: **Graph 1**, **Graph 2** and **Graph 3**.

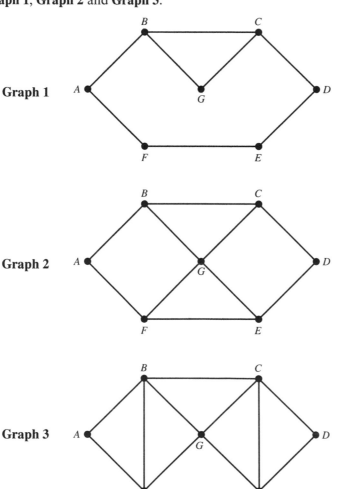

Graph 1

Graph 2

Graph 3

a) For **each** of the graphs, explain whether or not the graph is Eulerian.

b) The length of each edge connecting two vertices is 1 unit. Find, for **each** of the graphs, the length of an optimal 'Chinese postman' route, starting and finishing at A.   *(AQA, 2002)*

6 a) The network represents a road system in which the lengths of the roads are shown in kilometres. The road system has to be cleared of snow by a snow plough which is based at A. The snow plough only needs to travel along each road once in order to clear it. However, when all roads have been cleared it must return to its base at A.

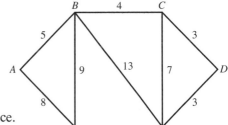

  i) State which roads the snow plough should drive along twice in order to travel the minimum total distance.
  ii) Hence solve the 'Chinese postman' problem for this network to find a route for the snow plough, starting and finishing at A.
  iii) The road BC becomes a dual carriageway and must be cleared twice, once in each direction. Redraw the network to take account of this and find, by inspection, a new route for the snow plough.

b) The original network has to be drawn as efficiently as possible by a pen operated by a computer. Explain how this can be done without lifting the pen from the paper and without tracing any of the edges more than once.   *(AQA, 1998)*

7 The network on the right represents the roads in a quiet housing estate. The only access to the estate is at A. A driving instructor uses the roads of this estate to provide practice for her less experienced pupils. She wishes to choose routes, starting and finishing at A, which involve as few repetitions of roads as possible.

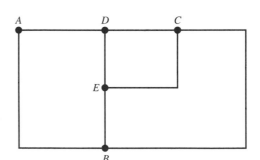

a) State why it is not possible to drive along all the roads exactly once, starting and finishing at A.
b) Describe a route along all the roads which involves as few repetitions of roads as possible.
c) Find a route with no right turns, which uses as many of the roads as possible, but involves as few repetitions of roads as possible.   *(AQA/NEAB, 1997)*

8 The diagram shows the time, in minutes, for a traffic warden to walk along a network of roads, where $x > 0$.

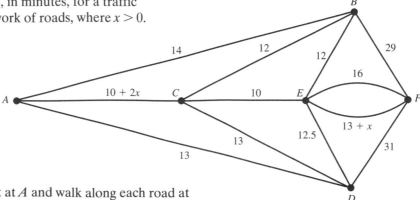

The traffic warden is to start at A and walk along each road at least once, before returning to A.

a) Explain why a section of roads from A to E has to be repeated.

b) The route ACE is the second shortest route connecting A to E. Find the range of possible values of x.

c) Find, in terms of x, an expression for the minimum distance that the traffic warden must walk and write down a possible route that he could take.

d) Starting at A, the traffic warden wants to get to F as quickly as possible. Use Dijkstra's algorithm to find, in terms of x, the minimum time for this journey, stating the route that he should take.

(AQA, 2001)

D1

5 The travelling salesman problem

This chapter will show you how to

♦ Identify a Hamiltonian cycle in relation to a tour round a network
♦ Convert a given network into the corresponding complete network of shortest distances
♦ Find possible solutions to the travelling salesman problem, using the Nearest Neighbour algorithm
♦ Provide an upper bound for the length of the optimal tour, using the Nearest Neighbour algorithm
♦ Find a lower bound for the optimal tour by using a minimum connector for a reduced network

5.1 The travelling salesman problem

Unlike the route inspection problem, where the aim was to travel along each edge of the network, the **travelling salesman problem** involves visiting every vertex.

Before you explore this problem in detail, you need to look at a little more graph theory.

> The problem is so named because it corresponds to a sales representative who needs to travel from base and attend appointments at a number of different locations before returning to base.

Hamiltonian cycles

You will recall that a **cycle** is a closed path. That is, a route within a graph which does not repeat any vertices and which finishes where it started. If a cycle visits **every vertex** it is called a **Hamiltonian cycle** or **Hamiltonian tour** (after the Irish mathematician Sir William Rowan Hamilton (1805–1865)). A graph which has a Hamiltonian cycle is said to be **Hamiltonian**.

In this diagram, *ACEGBDFA* is a Hamiltonian cycle. **Any** vertex of the graph could be taken as the start of this Hamiltonian cycle. For example, *GBDFACEG* is the same cycle.

Some graphs do not possess a Hamiltonian cycle. Any complete tour of this graph would need to visit the central vertex twice.

> In a cycle, you **cannot** repeat a vertex.

There is, as yet, no general theory for determining whether a given graph is Hamiltonian.

The classical travelling salesman problem

In the classical travelling salesman problem, the aim is to visit every vertex **once only**, and to return to the start, having travelled the least possible distance. In other words, you are trying to find a minimum-weight Hamiltonian cycle.

There are two ways in which a practical problem might not fit this classical pattern.

✦ The network may not possess a Hamiltonian cycle, as in Example 1.

D1

Example 1

Find the shortest tour of this network, starting and finishing at A.

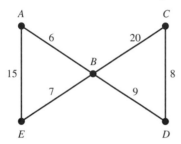

By inspection, you can see that the best route is clearly:

$ABDCDBEBA = 60$

Remember
A network is a graph with weighted edges.

Add the weights:
$AB + BD + \ldots$
$= 6 + 9 + 8 + 8 + 9 + 7$
$\quad + 7 + 6$
$= 60$

✦ Even if the network is Hamiltonian, the best route for the sales representative may not be a Hamiltonian cycle.

The only network for which the practical solution is bound to be a Hamiltonian cycle is one in which the shortest route between any two vertices does not pass through an intermediate vertex.

Example 2

Find the shortest tour of this network, starting and finishing at A.

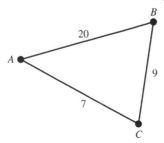

Although this network has a Hamiltonian cycle $ABCA$ with a total weight of 36, the best practical solution to the problem would be $ACBCA$, with a total weight of 32.

This means that the graph must be complete, so that there is an edge directly connecting each pair of vertices, and the weight of that edge must not be greater than the total weight for any indirect route between the vertices.

This is not the case in Example 2, where the direct route from A to B is of greater weight than the indirect route ACB.

$ABCA = 20 + 9 + 7 = 36$
$ACBCA = 7 + 9 + 9 + 7 = 32$

Converting to the classical problem

In all cases, you may, if necessary, draw a complete graph with weights corresponding to the shortest route between vertices (found by using Dijkstra's algorithm in complicated cases). Finding the minimum Hamiltonian cycle for this modified network will then give the best practical solution to the original problem.

To illustrate this, the network in Examples 2 will be re-examined.

Example 3

Redraw the network in Example 2 as a 'shortest routes' network. Find the solution to the classic travelling salesman problem for this modified network, and hence the solution to the practical problem.

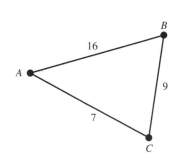

The shortest distances are $AB = 16$ (taking the route ACB), where $AC = 7$ and $BC = 9$. These give the network shown on the right.

The minimum Hamiltonian cycle for this network is $ABCA = 32$.

This gives the order in which to visit vertices in the original problem, taking the shortest route at each stage. The practical solution is, therefore, $ACBCA = 32$.

D1

The network in Example 1 is slightly more complex, but it can be modified, as shown in Example 4.

Example 4

Redraw the network in Example 1 as a 'shortest route' network. Find the solution to the classic travelling salesman problem for this modified network, and hence the solution to the practical problem.

Find the shortest route between each pair of vertices and draw the corresponding complete graph.

Check that you can see how the complete graph, shown on the right, corresponds to the original network.

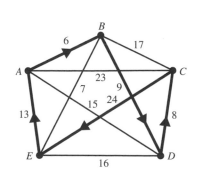

The minimum Hamiltonian cycle for this network (shown by the arrows in the diagram) is $ABDCEA = 60$.

This gives the order in which to visit the vertices in the original problem, taking the shortest route at each stage.

The practical solution is, therefore:

$ABDCDBEBA = 60$.

$6 + 9 + 8 + 24 + 13 = 60$

Finding the best tour

The only known method which guarantees to find the best route, or best **tour** in a given network is the brute-force method of checking every possible Hamiltonian cycle. Unfortunately, this is only feasible for very small networks, because of the large number of such cycles.

If the network has four vertices, A, B, C and D, then, starting from A, there is a choice of three possible vertices (B, C, D) to move to. Once there, there is a choice of two destinations. Finally, just the one route back to A. The number of possible tours is:

$3 \times 2 \times 1 = 3! = 6$.

Each route will appear twice, once in each direction. So, the actual number of different tours is $\frac{1}{2} \times 3! = 3$: that is, $ABCDA$, $ABDCA$ and $ACBDA$.

For five vertices, this becomes $\frac{1}{2} \times 4! = 12$ possible tours. Six vertices give $\frac{1}{2} \times 5! = 60$ possible tours.

In general, a network with n vertices has $\frac{1}{2} \times (n-1)!$ possible tours.

As a result, a fairly small network of, say, 16 vertices would require you to examine $\frac{1}{2} \times 15! = 6.54 \times 10^{11}$ different routes.

Hence, the usual methods employed are designed to find a good, though not necessarily optimal, solution in a reasonable length of time.

An algorithm which aims to find a satisfactory solution to a problem is called a **heuristic algorithm**.

> A **tour** is a route that visits every vertex and returns to the starting vertex.

D1

> A computer capable of examining a million routes every second would take about $7\frac{1}{2}$ days to solve this problem. If the network had 20 vertices, this time would rise to 1928 years!

The Nearest Neighbour algorithm

The simplest heuristic algorithm for the travelling salesman problem is the Nearest Neighbour algorithm. It can be stated as follows:

Step 1 Choose a starting vertex.

Step 2 From your current vertex, choose the edge with minimum weight, which leads to an unvisited vertex. Travel to that vertex.

Step 3 If there are unvisited vertices, go to Step 2. Otherwise, travel to the starting vertex.

If at any stage there are two or more equal edges to choose from, choose at random. It is then possible to rerun the algorithm, making the other choice(s) to ascertain if they give better solutions.

Remember that once a tour has been found, any vertex can be its starting point. The 'starting vertex' referred to in the algorithm is the starting vertex for the process of finding the tour – it does not have to be the actual start of the salesman's journey.

> This is another example of a greedy algorithm. That is, one which makes the immediately most advantageous choice at each stage without looking ahead to its implications.

As a result, you can repeat the Nearest Neighbour algorithm, taking each vertex in turn as the starting vertex. This often produces several different tours, from which you can select the best.

Example 5

Find the best tour for this network.

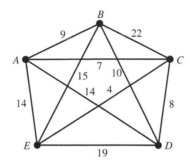

Applying the Nearest Neighbour algorithm with A as the starting vertex, you have:

From A, the 'nearest' vertex is C, with edge weight 7.
From C, the 'nearest' unvisited vertex is E, with edge weight 4.
From E, the 'nearest' unvisited vertex is B, with edge weight 15.
From B, the 'nearest' unvisited vertex is D, with edge weight 10.
There are now no unvisited vertices, so you complete the tour by travelling along DA, with edge weight 14.
The total weight for this tour, $ACEBDA$, is:

$$7 + 4 + 15 + 10 + 14 = 50$$

You now repeat the process with the other vertices as the starting vertex. You should check that you can see how these results come about:

With B as starting vertex, you have $BACEDB = 49$
With C as starting vertex, you have $CEABDC = 45$
With D as starting vertex, you have $DCEABD = 45$
With E as starting vertex, you have $ECDBAE = 45$

Notice that the last three of these results are, in fact, the same tour.

If the salesman's base were at A, you could conclude that the best tour available by using the Nearest Neighbour algorithm is $ABDCEA$, with a total weight of 45.

The other limitation of the Nearest Neighbour algorithm is that, unless the network is a complete graph, some or all of the starting vertices may not lead to a Hamiltonian cycle, even though such a cycle exists.

> In a **complete graph** there is an edge connecting every possible pair of vertices (see page 5).

D1

Example 6

Find the best tour for this network.

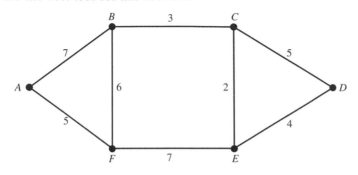

D1

If you apply the Nearest Neighbour algorithm, starting from A, it follows the route $AFBCED$, at which point you have no option but to repeat vertex C.

Try applying the algorithm with other starting vertices. You will find that, in every case, the process breaks down in that it is bound to repeat a vertex. Nevertheless, there is clearly a Hamiltonian cycle $ABCDEFA$ with a total weight of 31.

> This cycle goes round the perimeter of the network:
> $7 + 3 + 5 + 4 + 7 + 5 = 31$

Of course, you would generally convert the problem to a complete network of shortest routes, in which case the Nearest Neighbour algorithm would work. This is quite messy to draw and is better represented by a table. The Nearest Neighbour algorithm can then be carried out.

	A	B	C	D	E	F
A	–	7	10	15	12	5
B	7	–	3	8	5	6
C	10	3	–	5	2	9
D	15	8	5	–	4	11
E	12	5	2	4	–	7
F	5	6	9	11	7	–

Using A as the starting vertex, you have:

From A, the nearest unvisited vertex is F: $AF = 5$
From F, the nearest unvisited vertex is B: $FB = 6$
From B, the nearest unvisited vertex is C: $BC = 3$
From C, the nearest unvisited vertex is E: $CE = 2$
From E, the nearest unvisited vertex is D: $ED = 4$

All vertices have been visited, so return to A: $DA = 15$

The route is $AFBCEDA$, with a total weight = 35.

This corresponds to a practical solution consisting of $AFBCEDCBA$, because the route back from D to A has to pass through C and B.

You could now repeat the process using the other vertices as the starting point. You should try this for yourself, with B as the starting vertex, and confirm that this leads to a tour with a total weight of 32.

The other starting vertices give totals of 35 or 32. So the best solution available by this method is $BCEDFAB$ (corresponding to a practical solution of $BCEDEFAB$), with a total weight of 32.

Notice that in Example 6 the Nearest Neighbour algorithm does not find the Hamiltonian tour $ABCDEFA = 31$. This shows the limitations of the method.

Exercise 5A

1 The diagram on the right shows the direct routes, with distances in kilometres, between four towns.

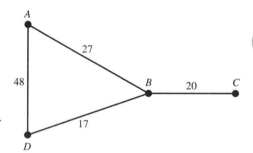

a) Draw a complete network with weights corresponding to the shortest distances between the four towns.

b) Taking A as the starting vertex, use the Nearest Neighbour algorithm on your network to find a possible tour. Find the total weight for this tour, and list the order in which the vertices would be visited on the original network.

2 a) Show that applying the Nearest Neighbour algorithm to the network shown on the right fails to find a Hamiltonian tour with three of the five possible starting vertices. List the tours produced, using the other two starting vertices, and state their total weights.

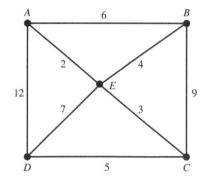

b) Draw a complete network, with weights corresponding to the shortest routes between the five vertices. Find the tour generated by the Nearest Neighbour algorithm, using A as the starting vertex. State its total weight and list the order in which the vertices would be visited on the original network.

3 A guide taking a party of tourists from a hotel to four attractions in a city estimates the walking times (in minutes) between the various locations, as shown in the table.

	Hotel	Museum	Art gallery	Cathedral	Guildhall
Hotel	–	5	8	10	4
Museum	5	–	6	7	2
Art gallery	8	6	–	3	5
Cathedral	10	7	3	–	8
Guildhall	4	2	5	8	–

Using the Nearest Neighbour algorithm with various starting vertices, find the route which you would recommend to the guide.

4 The table shows the shortest routes (in km) between six Somerset towns.

	Cheddar	Frome	Glastonbury	Radstock	Shepton Mallet	Wells
Cheddar	–	38	20	26	24	13
Frome	38	–	34	12	20	27
Glastonbury	20	34	–	32	15	14
Radstock	26	12	32	–	16	20
Shepton Mallet	24	20	15	16	–	10
Wells	13	27	14	20	10	–

D1

A lorry needs to travel from a warehouse in Frome and make deliveries to shops in each of the other five towns.

a) Use the Nearest Neighbour algorithm, with Frome as the starting vertex, to find a possible route, and state its length.

b) Show that there is a shorter route if Radstock is used as the starting vertex. Hence, list the order in which the lorry should visit the towns.

5 The table shows the direct distances (in miles) between four locations. There are no other direct links between them.

	A	B	C	D
A	–	5	–	6
B	5	–	–	7
C	–	–	–	3
D	6	7	3	–

a) Fill in the blank cells in the table with the shortest indirect routes available.

b) Use the Nearest Neighbour algorithm, with A as the starting vertex, to find a Hamiltonian tour of your completed network. State the total weight and list the order in which this tour would actually pass through the locations.

6 An orchestra and chorus is available as a whole unit (option A), or as four subgroups (B, C, D and E) offering smaller scale performances. The organiser of a six-day music festival wants to have a grand start and finish with option A, but to have the other four options appearing on the remaining four days. Because of travelling expenses etc, as performers come and go, the changeover costs between the various options vary, according to this table.

	A	B	C	D	E
A	–	250	300	150	400
B	250	–	120	360	220
C	300	120	–	260	170
D	150	360	260	–	290
E	400	220	170	290	–

Advise the organiser as to the least expensive order in which to schedule the options.

5.2 Upper and lower bounds

When you find a solution to the travelling salesman problem, you do not usually know whether it is the best possible (optimal) solution. It would be useful to have some idea of how close it might be to the optimal solution.

Suppose, for example, that you could show that the optimal solution to your problem had a total weight lying between 250 and 280. If you subsequently found a solution with a total weight of 254, you would know that this was either optimal or very close, and that it would probably not be worth the effort of trying to improve on it.

The 280 in this example is an **upper bound**. The optimal solution cannot be higher than this.

The 250 is a **lower bound**. The optimal solution cannot be lower than this.

Finding upper and lower bounds

This is best explained by means of an example.

Example 7

Find a) an upper bound and b) a lower bound for W, the total weight of the optimal tour for the network shown below. Hence, state the inequalities which W must satisfy.

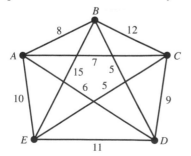

a) Finding an upper bound is straightforward. The total weight of any tour provides you with an upper bound. The usual approach is to apply the Nearest Neighbour algorithm with all possible starting vertices. The best result you obtain is used as an upper bound.

Using A as the starting vertex, you obtain:
$ADBCEA = 6 + 5 + 12 + 5 + 10 = 38$

Using B as the starting vertex, you obtain
$BDACEB = 5 + 6 + 7 + 5 + 15 = 38$

Using C as the starting vertex, you obtain
$CEADBC = 5 + 10 + 6 + 5 + 12 = 38$

Using D as the starting vertex, you obtain
$DBACED = 5 + 8 + 7 + 5 + 11 = 36$

Using E as the starting vertex, you obtain
$ECADBE = 5 + 7 + 6 + 5 + 15 = 38$

Hence, the best upper bound from this method is 36.

D1

b) Separate the edges connected to A from the rest of the network, giving the two sub-graphs shown below.

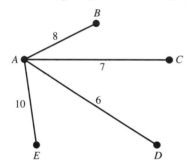 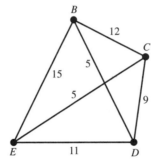

> You **delete** the vertex A from the network.

Any tour must enter and leave A along two of the edges in the left-hand diagram. This cannot be achieved in less than $6 + 7 = 13$.

Any tour must connect B, C, D and E together. This cannot be achieved in less than the minimum connector of the network in the right-hand diagram. Using either Prim's algorithm or Kruskal's algorithm (see pages 11 to 13), you can see that this is BD, CD and CE, with a total weight of $5 + 9 + 5 = 19$.

Putting these two facts together, you can see that the optimal tour cannot be shorter than $13 + 19 = 32$. This is a possible lower bound for the optimal solution.

You may be able to improve on this lower bound by separating one of the other vertices from the network. Separating B, you have:

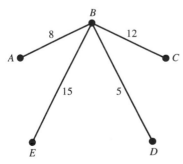 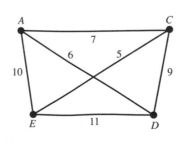

The shortest route into and out of B is $5 + 8 = 13$

The minimum connector of $ACDE$ is $5 + 7 + 6 = 18$

The lower bound obtained from these is $13 + 18 = 31$

As you already know, the optimal tour cannot be less than 32. Hence, this new lower bound is not useful.

Separating C, you have:

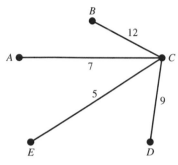
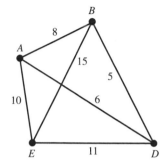

The shortest route into and out of C is $5 + 7 = 12$

The minimum connector of $ABDE$ is $5 + 6 + 10 = 21$

The lower bound obtained from these is $12 + 21 = 33$. This is an improved lower bound because you previously knew that the optimal tour could not be less than 32, but you now know that it cannot be less than 33.

Separating D, you have:

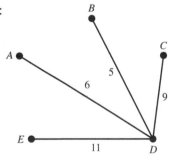
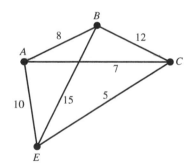

The shortest route into and out of D is $5 + 6 = 11$

The minimum connector of $ABCE$ is $5 + 7 + 8 = 20$

The lower bound obtained from these is $11 + 20 = 31$. So, this is not an improvement.

Separating E, you have:

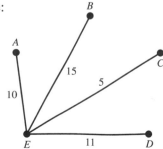
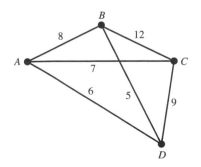

The shortest route into and out of E is $5 + 10 = 15$

The minimum connector of $ABCD$ is $5 + 6 + 7 = 18$

The lower bound obtained from these is $15 + 18 = 33$

This value of the lower bound is the same as that obtained from separating C. Hence, the best lower bound available from this method is 33.

As a result, you can state that the total weight, W, of the optimal tour must satisfy the inequality:

$$33 \leqslant W \leqslant 36$$

D1 The process for finding a lower bound is:

Step 1   Choose a vertex. Separate (delete) this vertex and the edges connected to it from the remainder of the network.

Step 2   Find the length of the minimum connector for the remaining network.

Step 3   Add to this minimum connector the two shortest edges entering the chosen vertex. Record this total.

Step 4   Repeat Steps 2 and 3 for each vertex in turn.

Step 5   Take as the lower bound the largest of the recorded totals.

> **Note**   Sometimes the minimum connector and the two shortest edges form a Hamiltonian cycle. In this case, what you have is a possible tour (and therefore an upper bound) which is also a lower bound. This means that this tour is the optimal tour.

Often, examination questions only require Steps 1 to 3 to find a typical lower bound by deleting a specified vertex. Similarly, a typical upper bound can be asked for by starting with a specified vertex.

Example 8

The table shows the distances in miles between five locations.

a) Use the Nearest Neighbour algorithm with starting vertex A to find an upper bound for the optimal tour.

b) Delete vertex A from the table. Use Prim's algorithm to find the minimum connector for the remaining network. Hence, find a lower bound for the optimal tour, and state the inequalities which the optimal tour must satisfy.

	A	B	C	D	E
A	–	14	22	18	12
B	14	–	15	32	25
C	22	15	–	20	13
D	18	32	20	–	28
E	12	25	13	28	–

a) Proceed as follows,

From A, the nearest unvisited vertex is E, $AE = 12$
From E, the nearest unvisited vertex is C, $EC = 13$
From C, the nearest unvisited vertex is B, $CB = 15$
From B, the nearest unvisited vertex is D, $BD = 32$

All vertices have now been visited, so return to A: $DA = 18$

The route is $AECBDA$, with total weight $= 90$

Hence, the optimal tour is 90 miles or less. The required upper bound is 90 miles.

b) Disregard vertex A.

	A	1	2	3	4
		B	C	D	E
A	–	14	22	18	12
B	14	–	15	32	25
C	22	(15)	–	20	13
D	18	32	(20)	–	28
E	12	25	(13)	28	–

Taking B as the starting vertex for the remaining network and applying Prim's algorithm, as shown, you obtain the minimum connector of $BCDE$ as BC, CD and CE, with a total weight of 48.

The two shortest edges connecting to A are AE and AB, with a total weight of 26.

A lower bound for the optimal tour is, therefore:

$48 + 26 = 74$ miles

Following from parts a) and b), you can state:

$74 \leqslant$ Optimal tour $\leqslant 90$

> In this case, you are given the starting vertex, A, so there is no need to consider any other starting vertex for the optimal tour.

Exercise 5B

1 a) By separating and deleting vertex A from the network on the right, obtain a lower bound for the optimal tour. Explain, by reference to the network, why a tour of this length is not possible.

  b) By separating and deleting vertex B, obtain another lower bound for the optimal tour and explain why this is, in fact, the length of the optimal tour.

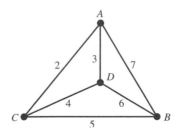

2 The diagram shows the cost, in dollars, of flying between five locations in the United States. A tourist arrives in and leaves the country at A, and wishes to travel to each of the locations in turn.

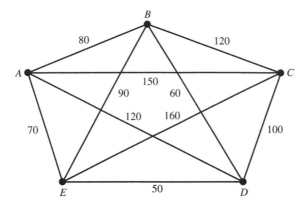

D1

a) Apply the Nearest Neighbour algorithm, taking each vertex in turn as the starting vertex, to obtain the best available upper bound for the cost of the least expensive route the tourist could take.

b) By deleting each vertex in turn, obtain the best available lower bound.

c) State the inequalities satisfied by the cost of the optimal route.

3 The table shows the direct distances, in kilometres, between five Midland towns. A sales representative is to travel to each town in turn, starting and finishing in Coventry.

	Coventry	Nottingham	Leicester	Derby	Stoke-on-Trent
Coventry	–	77	37	12	93
Nottingham	77	–	40	24	80
Leicester	37	40	–	45	82
Derby	12	24	45	–	56
Stoke-on-Trent	93	80	82	56	–

a) Using Coventry as the starting vertex and applying the Nearest Neighbour algorithm, find an upper bound for the length of the optimal tour.

b) By deleting Coventry from the network and applying Prim's algorithm to the remainder, find a lower bound for the length of the optimal tour.

4 a) Draw a table depicting the shortest route between the vertices of the network shown on the right.

b) Show that the best tour available by using the Nearest Neighbour algorithm exceeds the optimal tour by no more than 2.

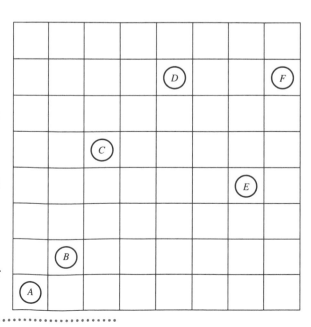

5 In an orienteering competition, contestants must visit each of five checkpoints once only and return to the start. They gain points depending on the difficulty of the route they follow – the harder the route, the more points (but the greater the chance that they will not complete the course in the required time). The table shows the points to be gained on the various routes.

	Start	A	B	C	D	E
Start	–	10	8	14	9	12
A	10	–	5	6	18	15
B	8	5	–	10	12	7
C	14	6	10	–	15	6
D	9	18	12	15	–	10
E	12	15	7	6	10	–

a) George wishes to travel the easiest route (least total points). Find the best lower and upper bounds for his optimal total, and state the best route that you can find for George to follow.

b) Shahidar plans to gain as many points as possible (the hardest route). By adapting the Nearest Neighbour algorithm and Prim's algorithm, find the best lower and upper bounds for her optimal total, and state the best route that you can find for Shahidar to follow.

6 A knight's move in chess consists of moving two squares parallel to one side of the board and then one square at right angles to this.

Belinda and Manuel play a game in which one of them places counters on six squares of the board and the other must make a 'knight's tour', starting and ending on one of the marked squares and 'capturing' each of the other counters en route, making as few moves as possible.

The diagram shows the positions of the counters that Belinda has set up.

a) Draw up a table to show the numbers of moves required to travel between each of these positions.

b) Find upper and lower bounds for the number of moves that Manuel will have to make.

D1

Summary

You should know how to ...

1 Recognise a **Hamiltonian cycle** or **Hamiltonian tour**, which is a closed path visiting every vertex of the graph once only.

2 Convert problems to the classic **travelling salesman problem** by drawing a complete network with weights equal to the shortest distances between vertices in the original network.

3 Apply the **Nearest Neighbour algorithm** to find a possible Hamiltonian tour. The algorithm is:

    Step 1 Choose a starting vertex.

    Step 2 From your current vertex, choose the edge with minimum weight, which leads to an unvisited vertex. Travel to that vertex.

    Step 3 If there are unvisited vertices, go to Step 2. Otherwise travel to the starting vertex.

If at any stage there are two or more equal edges to choose from, choose at random.

4 Use the Nearest Neighbour algorithm to find an **upper bound** for the optimal tour.

5 Find a **lower bound** for the optimal tour, using the minimum connector for a reduced network, as follows.

    Step 1 Choose a vertex. Separate this vertex and the edges connected to it from the remainder of the network.

    Step 2 Find the length of the minimum connector for the remaining network.

    Step 3 Add to this minimum connector the two shortest edges entering the chosen vertex. Record this total.

The total recorded is a lower bound. If the process is repeated for each vertex, the largest total recorded is the best lower bound.

D1

Revision exercise 5

1 A country policeman wants to start at the police station P and visit each of the villages Q, R, S, T, U, V once before returning to the station. The distances (in miles) by the easiest routes are shown in the table.

	P	Q	R	S	T	U	V
P	–	5	4	3	4	5	4
Q	5	–	7	3	4	5	8
R	4	7	–	6	8	6	7
S	3	3	6	–	4	6	9
T	4	4	8	4	–	7	8
U	5	5	6	6	7	–	8
V	4	8	7	9	8	8	–

a) Use the Nearest Neighbour algorithm, starting from P, to find one possible route for the policeman.

b) Find the length of a minimum connector of the villages Q, R, S, T, U and V.

c) Deduce that the length, L, of the shortest route for the policeman satisfies:

$$32 \leqslant L \leqslant 34$$

(AQA, 2002)

2 A researcher starts at the desk, D, of the British Library and wishes to consult books at locations E, F, G, H, I and J in the library. The time, in minutes, needed to walk between any two places is given in the table.

	D	E	F	G	H	I	J
D	–	7	6	9	7	10	8
E	7	–	11	6	12	10	7
F	6	11	–	10	11	11	12
G	9	6	10	–	13	11	8
H	7	12	11	13	–	8	12
I	10	10	11	11	8	–	9
J	8	7	12	8	12	9	–

D1

a) Use the Nearest Neighbour algorithm, starting at D, to find one appropriate order in which the researcher might visit the six locations before returning to D. State the total time needed for walking around this particular route.

b) Find a minimum connector of just the six locations E, F, G, H, I and J, and state its total length. Draw a tree representing your minimum connector.

c) Explain why, in this case, the route which you found in part a) is definitely the shortest possible.

(AQA, 2002)

3 A family is staying in Dublin and is to visit four places of interest in the city. There is a one-way traffic system in the city and consequently the time taken to travel from A to B is different from the time taken to travel from B to A.

The following diagram shows the four places to be visited, together with the time taken, in minutes, to travel between each two places.

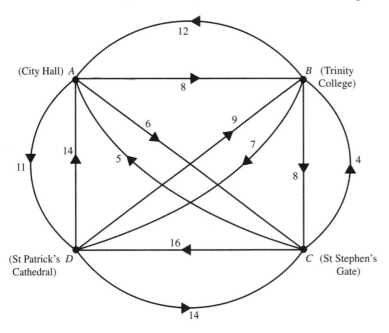

a) The family starts the tour at A and visits each of the other three places once before returning to A.
   i) Use the Nearest Neighbour algorithm to find an upper bound for the travelling time of the tour.
   ii) Find the number of different possible tours.

b) Write down an expression for the number of possible tours if the family were to start at A and visit each of n other places once before returning to A.

(AQA, 2003)

4 The network **N** is illustrated below.

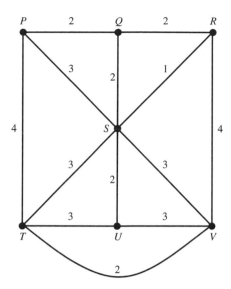

a) Use the 'Chinese postman' algorithm to find the length of a shortest closed walk in **N** which uses all its edges.

b) Apply the Nearest Neighbour algorithm, starting and finishing at S, to construct a Hamiltonian cycle of **N**. State its length.

(In this case, the algorithm works even though the network is not complete.)

c) i) By considering the vertex U, explain why every Hamiltonian cycle of **N** uses at least one edge of length 3.

   ii) By considering **N** with edges PT and RV deleted, explain briefly why each Hamiltonian cycle of **N** uses at least one edge of length 4.

   iii) Show that the Hamiltonian cycle which you constructed in part b) is the shortest possible in **N**.

d) The network **N** represents seven computer centres
P, Q, R, S, T, U, V and the direct cable links between them.
The number on each edge shows the time, in seconds, that it
takes for a piece of information to pass along that link.
Whenever a piece of information is received at one centre and
then passed on to another, there is a further 20 seconds delay.

  i)   A piece of information has to be sent from S to another
       centre and then passed on to another, and then passed
       on again, until it is eventually sent back to S having
       been received at least once at each of the other
       computer centres.
       Use your earlier answers to find the minimum time that
       this will take.

  ii) A technician based at S wants to check every link in the
       system by sending a piece of information from S to
       another centre and then on to another and then on again,
       until it is eventually sent back to S having used **each link**
       at least once.
       Use your earlier answers to find the minimum time
       that this will take.

(AQA, 2003)

5 A machine is used for producing ice cream in six flavours. The
   machine produces one flavour of ice cream at a time. It needs to be
   cleaned before changing flavours. The times taken to clean the
   machine depend on the two flavours involved and these times, in
   minutes, are given in the table. The machine is set to produce each
   flavour in sequence before repeating the cycle. The machine can
   start the cycle with any flavour.

From \ To	Vanilla	Lemon	Orange	Raspberry	Coffee	Mint
Vanilla	–	40	35	35	42	40
Lemon	30	–	25	26	30	45
Orange	20	45	–	30	35	34
Raspberry	35	40	30	–	40	25
Coffee	25	35	22	30	–	30
Mint	30	40	34	25	35	–

a) Upper bound times in excess of three hours are produced by using the Nearest Neighbour algorithm when starting with vanilla or raspberry. Use the same method to find **four** further upper bounds for the total cleaning time, showing that only one produces a total time of under 175 minutes.

b) The manager of the factory realises that he must select one number from each row and one number from each column to represent a complete cycle. By using a 'greedy' approach, find a cycle that will produce a total cleaning time of under 170 minutes.

(AQA, 1998)

6 It is the year 2302 and space travel between the planets of the solar system has become commonplace.

Sarah, who is based on Mars, is planning a tour of four other planets before returning to Mars. The planets orbit the Sun, and the distance between each pair of planets varies. The following table gives the distance between planets, in space units, in terms of a variable x.

D1

	Mars (M)	Jupiter (J)	Saturn (S)	Neptune (N)	Pluto (P)
Mars (M)	–	$3x + 6$	$x + 24$	$2x + 17$	$5x$
Jupiter (J)	$3x + 6$	–	$x + 4$	$2x - 3$	$3x - 2$
Saturn (S)	$x + 24$	$x + 4$	–	$x + 13$	$2x - 1$
Neptune (N)	$2x + 17$	$2x - 3$	$x + 13$	–	$x + 6$
Pluto (P)	$5x$	$3x - 2$	$2x - 1$	$x + 6$	–

a) Given that Jupiter is the nearest planet to Mars, find three inequalities in x.

Sarah plans a tour using the Nearest Neighbour algorithm and finds that, for one particular **integer** value of x, this algorithm gives a **unique** tour of M, J, S, P, N, M.

b) By considering the fact that Sarah's tour visits Saturn immediately after Jupiter, find two further inequalities in x.

c) Hence, find the value of the integer x, and the total length of Sarah's tour.

(AQA, 2002)

D1

7 G_n is the graph with n vertices labelled 1, 2, 3, ..., n and with two vertices joined by an edge if the sum of their labels is odd. So, for example, G_4 and G_5 are as drawn below.

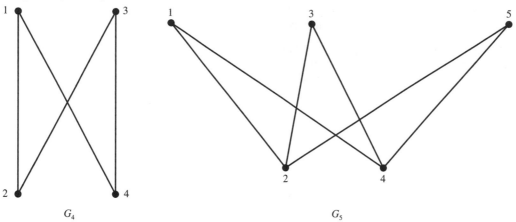

a) Draw G_6.

b) i) Give an example of a Hamiltonian cycle in G_6.

  ii) For what values of n is G_n Hamiltonian?

c) If n is even, what, in terms of n, is the degree of each vertex of G_n?

d) For what values of n is G_n Eulerian? Justify your answer. *(AQA, 2004)*

6 Matching problems

This chapter will show you how to

◆ Model matching problems as bipartite graphs
◆ Find an alternating path through a partially matched graph
◆ Use the maximum matching algorithm to improve on a known matching or to establish that the known matching is maximal

6.1 Bipartite graphs

In this chapter, you will examine the problem of matching up the members of two sets. For example, you may have five workers available and five jobs to be done, but each of the workers may have the skills for only some of the jobs. The problem is to find a way of pairing up workers and jobs so that all the jobs are done, and done by qualified workers.

Once again, you model the situation as a graph and so you need to look at the relevant theory.

- The vertices of the graph represent the two sets.
- Each edge joins a member of the first set to a member of the second set.

> In this example, the two sets are the workers and the jobs.

Such a graph is called a **bipartite graph**.

In a bipartite graph, the numbers of vertices in the two sets need not be equal and some of the possible connections may not exist. If every vertex in the first set is joined to every vertex in the second set, you have a **complete bipartite graph**. A complete bipartite graph connecting m vertices to n vertices is labelled $K_{m,n}$.

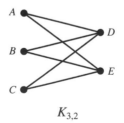

$K_{3,2}$

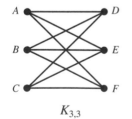

$K_{3,3}$

> The edges only join vertices in one set to vertices in the other set. There are no edges between vertices of the same set.

The adjacency matrix for a bipartite graph takes a typical form (provided that the vertices are listed with one set followed by the other). For example, consider the following graph and its matrix.

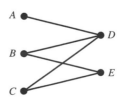

	A	B	C	D	E
A	0	0	0	1	0
B	0	0	0	1	1
C	0	0	0	1	1
D	1	1	1	0	0
E	0	1	1	0	0

D1

There are two square blocks of zeros in this matrix, corresponding to the two sets $\{A, B, C\}$ and $\{D, E\}$. Because of this, it is usual to write the adjacency matrix for bipartite graphs in a more compact form, as shown on the right.

	D	E
A	1	0
B	1	1
C	1	1

If an adjacency matrix does not list the vertices in the appropriate order, it is less clear that you are dealing with a bipartite graph.

D1

Example 1

Show that the following matrix represents a bipartite graph.

	A	B	C	D	E	F
A	0	0	1	0	0	0
B	0	0	0	1	1	0
C	1	0	0	1	1	0
D	0	1	1	0	0	1
E	0	1	1	0	0	1
F	0	0	0	1	1	0

. . . .

The graph has seven edges: AC, BD, BE, CD, CE, DF and EF.

To decide whether the graph is bipartite, try to allocate the vertices to two sets, X and Y.

> Each edge is denoted by 1 in the matrix.

    If A is in X, then C must be in Y, as there is an edge AC.
    If C is in Y, then D and E must be in X, as there are edges CD and CE.
    If D is in X, then B and F must be in Y, as there are edges DB and DF.

You can now check that there are no edges connecting members of X together or members of Y together, or you can redraw the matrix with the two sets of vertices separated, as shown on the right.

	A	D	E	B	C	F
A	0	0	0	0	1	0
D	0	0	0	1	1	1
E	0	0	0	1	1	1
B	0	1	1	0	0	0
C	1	1	1	0	0	0
F	0	1	1	0	0	0

You can now see the two tell-tale blocks of zeros, indicating that this is indeed a bipartite graph. You can draw the matrix in the more usual compact form, as shown on the right.

	B	C	F
A	0	1	0
D	1	1	1
E	1	1	1

Modelling a problem as a bipartite graph

Example 2

A school wishes to run three language classes, one in each of French, German and Spanish. There are four language teachers available – Polly Glott, who can speak all three languages, Lynne Gwist, who speaks German and Spanish, Frank O'File, who speaks just French, and Dick Shonary, who speaks French and Spanish. Illustrate this information using a bipartite graph.

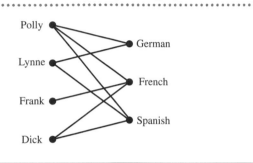

The number of teachers exceeds the number of classes, so one of them will not be used. There are several ways in which the allocation could be made. For example, Lynne takes German, Frank takes French and Dick takes Spanish.

D1

The information in Example 2 could have been expressed as shown below.

	German	French	Spanish
Polly	✓	✓	✓
Lynne	✓		✓
Frank		✓	
Dick		✓	✓

This table is effectively an **adjacency matrix** for the graph.

Matchings

For a bipartite graph, a **matching**, M, is a subset of the edges of the graph with the property that no two edges in M share a vertex.

It is usual to show a matching by drawing the edges in M using thicker lines. Here are some examples of matchings for a particular graph.

A matching 'pairs up' vertices in the two sets.

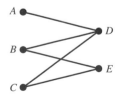

M possesses no edges.

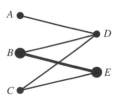

M has one edge, BE.

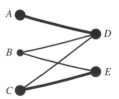

M has two edges, and is a **maximal matching**.

In a **maximal matching**, it is not possible to include more edges in M without having two edges sharing a vertex.

In some cases, the maximal matching includes all the vertices in the smaller set, but this is not always the case. Consider the graph below.

Vertices E and F are both connected only to C, and one of these vertices must remain unconnected.

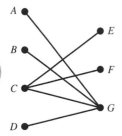

The maximal matching can have only two edges. For example:

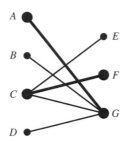

For a maximal matching to include every vertex in the smaller set the following conditions must be met.

✦ Every pair of vertices in the smaller set must between them connect to at least two vertices in the larger set.

✦ Every set of three of vertices in the smaller set must between them connect to at least three vertices in the larger set.

And so on.

If the two sets in your graph have the same number of vertices, and a matching exists which includes every vertex, you have a **complete matching**. For example:

> This is known as Hall's Marriage Theorem, because matching problems were originally stated in terms of trying to marry off a set of men to a set of women.

This graph possesses a complete matching.

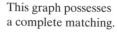

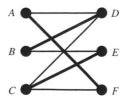

Here is another example.

Example 3

The table shows a group of four workers, John, Kamlesh, Lara and Meera, and a set of four tasks – A, B, C and D. The ticks indicate information as to which worker is qualified for which tasks. Illustrate this using a bipartite graph and investigate whether it is possible to pair up each worker with a task for which she/he is qualified.

	A	B	C	D
J		✓		
K	✓	✓		
L	✓	✓	✓	✓
M	✓	✓		

The table is an adjacency matrix and the diagram on the right shows the corresponding bipartite graph.

If you try to assign a worker to each task, a problem arises immediately. Both task C and task D can only be done by Lara, so it is not possible to assign both of these tasks.

The best that can be done is to match three people with three tasks. This can be done in several ways, one of which is illustrated on the right.

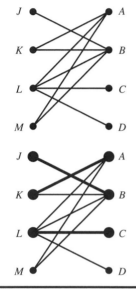

D1

Example 4

Show that there is a complete matching for the graph shown on the right.

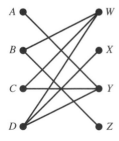

There are three edges which are forced on you by the fact that some vertices have only the one connecting edge, namely (A, Y), (B, Z) and (D, X). This means that the remaining pairing must be (C, W).

Hence, a complete matching exists, as shown on the right. It is the only one possible, as you had no choice at any stage.

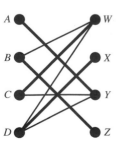

Exercise 6A

1 A school music teacher asks pupils to put themselves forward to play in a group. The group will consist of four people: a drummer, a bassist, a guitarist and a vocalist. Five pupils come forward, and the parts they could take are shown in the table.

Show this information as a bipartite graph.

	Drums	Bass	Guitar	Vocals
Ali	✓	✓		
Ben		✓	✓	✓
Cass	✓			✓
Dee		✓	✓	
Eve			✓	✓

2 Illustrate all possible maximal matchings for this bipartite graph.

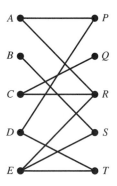

3 Find the two possible complete matchings for this bipartite graph.

4 At a dog show, there are five breed classes, each of which requires a judge, and five judges are available with knowledge of some of those breeds. Their expertise is summarised in the table.

a) Show this information as a bipartite graph.

b) Find a complete matching, explaining your working. Is it the only possible complete matching?

Judge	Breeds
Mr Lee	Corgi, Beagle
Mrs Tweed	Samoyed, Corgi
Mrs Pinkham	Beagle, Whippet
Mr Zapata	Corgi
Miss Floyd	Corgi, Whippet, Leonberger

5 The owner of a row of four houses proposes to paint each of the front doors a different colour. She gives the tenants the chance to say which of the five possible colours they would definitely **not** want. This information is summarised in the table at the top of page 87.

a) Draw a bipartite graph to show which colours could go with which houses.

b) Find a possible maximal matching.

	House number			
	1	2	3	4
Blue		✗	✗	
Red		✗		✗
Green	✗		✗	✗
White	✗			✗
Yellow	✗	✗	✗	

6.2 Improving a matching

In the examples considered so far, it is quite easy to see what the best matching is for a given bipartite graph. However, in more complex examples, it can be quite difficult to obtain the maximal matching.

It is, of course, easy to find a partial matching of some sort. Just selecting one edge at random gives a partial matching, although it is usually possible to do much better than that. You then need an algorithm which will:

◆ improve on a known matching if this is possible, and
◆ indicate when you have found a maximal matching.

Consider the example on the right. You have the bipartite graph, for which a partial matching has been found, indicated by the thick edges. Call the set of edges in this matching M.

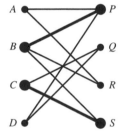

Starting with an unmatched vertex in the left-hand set, say A, you will look for a path through the graph, travelling **left to right on edges that are not in M** and **right to left on edges that are in M**. You hope that the final vertex visited will be an unmatched vertex. If so, the path is called an **alternating path** and you are said to have made a **breakthrough**.

◆ An **alternating path** joins an unmatched vertex in the left-hand set to an unmatched vertex in the right-hand set by edges which are alternately in and not in the set, M, of matched edges.

In this case, you have the alternating path $A - P - B - S - C - Q$. Illustrated separately, it is:

Notice that, in an alternating path, the number of edges which are not in M is bound to be one more than the number which are in M.

You now **change the status** of these edges, so that edges which are in the matched set are removed from it, and edges which are not in the matched set are added to it. The path then looks like this.

You now have a new matching for the graph, with one more edge than before, as shown.

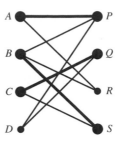

You now repeat the process, using D as the starting vertex, in the hope of improving the matching further.

There is an alternating path starting at D: namely, $D - P - A - R$. So, you again have a breakthrough.

Changing the status of these edges gives:

D1

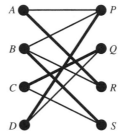

The resulting new matching is a complete matching, because every vertex is now connected, as shown.

You can formalise the above process, variously called the **maximum matching algorithm**, the **alternating path algorithm** or the **matching improvement algorithm**, as follows.

Step 1   Find a matching, M. (This could just be one edge, but it is usual to start with as good a matching as can quickly be found.)

Step 2   Search for an alternating path. If none exists, go to Step 4.

Step 3   Form a new matching by changing the status of the edges in the alternating path, and go to Step 2.

Step 4   Stop. The current matching is maximal.

Finding an alternating path

The systematic search for an alternating path can be quite difficult, as some starting vertices may lead to several possible routes.

Consider the graph on the right.

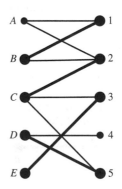

You can search for an alternating path, starting from A, by drawing a tree diagram.

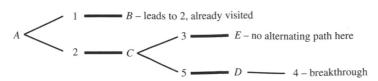

However, there are other approaches, which are useful in more complex cases. You will be shown two alternatives. There is no need for you to know both of these – you should learn the method with which you feel more comfortable.

Method 1

Step 1 Choose an unmatched vertex in the left-hand set. Label it *.

Step 2 For each of the most recently labelled left-hand vertices, identify all unlabelled right-hand vertices to which it is connected. If no such vertex exists, go to Step 8.

Step 3 Label these with the name of the connected left-hand vertex.

Step 4 If any of the vertices labelled in Step 3 is unmatched, you have a breakthrough. Go to Step 9.

Step 5 For each of the most recently labelled right-hand vertices, identify any unlabelled left-hand vertex to which it is connected by an edge in the matched set, M. If no such vertex exists, go to Step 8.

Step 6 Label these with the name of the connected right-hand vertex.

Step 7 Go to Step 2.

Step 8 If there are unlabelled unmatched vertices in the left-hand set, go to Step 1. Otherwise stop – the matching is maximal

Step 9 Identify the alternating path by tracing back through the labels.

To illustrate Method 1, apply it to the last graph on page 88.

the last graph on page 88.

Example 5

Find an alternating path for the graph shown on the right. Hence, find a complete matching for the graph.

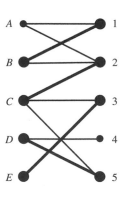

· ·

Label vertex A with an asterisk * and vertices 1 and 2 connected to it with A.

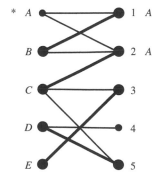

Neither of the right-hand vertices just labelled is unmatched. So, look for left-hand vertices to which they are connected by matched edges:

> 1 connects to B, so label B with 1.
> 2 connects to C, so label C with 2.

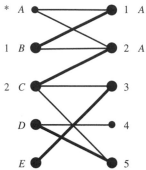

1–B and 2–C are matched edges.

Now look for right-hand vertices to which B and C are connected. There are no unlabelled vertices connected to B, but C connects to 3 and 5, so label these with C.

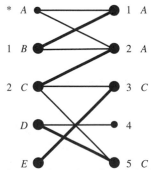

B connects to 1 and 2, both labelled A. C connects to 3 and 5, which are not labelled yet.

Neither of the right-hand vertices just labelled is unmatched. Look for left-hand vertices to which they are connected by matched edges:

> 3 connects to E, so label E with 3.
> 5 connects to D, so label D with 5.

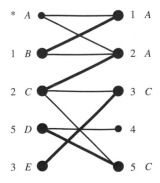

Now look for right-hand vertices to which D and E are connected. There are no unlabelled vertices which connect to E, but D connects to 4, so label this with D.

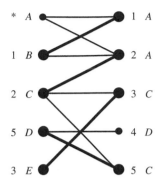

Vertex 4 is an unmatched vertex, so you have a breakthrough. You can now trace back to find the alternating path:

$$4 - D - 5 - C - 2 - A.$$

You can now change the status of the edges on this alternating path, resulting in the matching shown below.

> The alternating path follows an unmatched, a matched, an unmatched, a matched, and an unmatched edge.

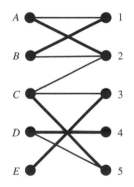

All vertices are now matched, so you have a complete matching.

Method 2

Turn the graph into a directed network. Put arrows from right to left on all the edges in the matched set M, and from left to right on the edges not in M. Give each edge a weight of 1. The algorithm is as follows.

Step 1 Choose an unmatched vertex in the left-hand set and give it the permanent label 0.

Step 2 Apply Dijkstra's algorithm until either an unmatched right vertex is reached (that is, breakthrough) or no further labelling is possible.

Step 3 If an unmatched right vertex is reached, trace the path back and stop. You have found an alternating path.

Step 4 If any unmatched left vertices remain, go to Step 1. Otherwise stop – the matching is maximal.

Example 6

Find an alternating path for the graph on the right. (This is the same graph as in Example 5.)

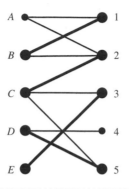

D1

First, redraw the graph as a directed network, as shown.

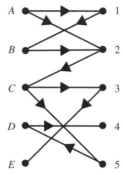

> Arrows go right to left on **matched** edges 1B, 2C, 3E, 5D and left to right on the other edges A1, A2, B2, C3, C5, D4.

Take the unmatched vertex, A, as the start vertex. All edges have weight 1. After one stage of Dijkstra's algorithm you have:

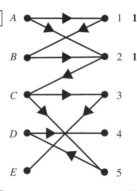

After three iterations of the algorithm, you have:

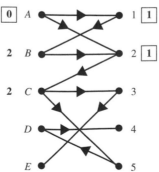

> Vertices B and C can each be reached from vertex A in two stages.

Continuing in this way, you have after five iterations:

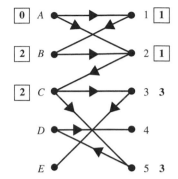

Vertices 3 and 5 can each be reached from vertex *A* in three stages.

After seven iterations, you have:

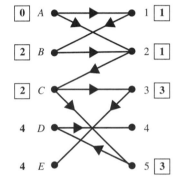

Vertices D and E can be reached from vertex A in four stages.

After ten iterations, you have:

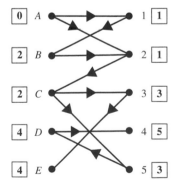

You have now reached vertex D, which is unmatched, so you have a breakthrough. Now trace back through the route $4 - D - 5 - C - 2 - A$, which is an alternating path.

D1

Exercise 6B

1 Each of these diagrams shows a partial matching for a bipartite graph. Using the maximum matching algorithm, decide whether the given matching is maximal. If it is not, find a maximal matching.

a)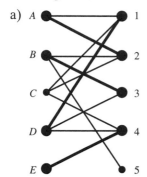

b)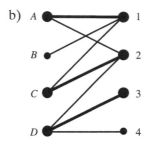

c)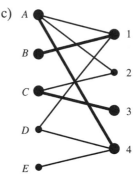

d)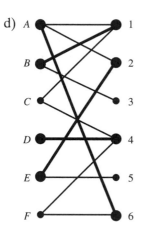

e)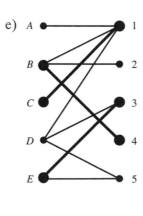

2 Four workers, A, B, C and D, are qualified for some of the tasks 1, 2, 3 and 4, as follows.

Worker	A	B	C	D
Task	1, 3	1, 2	2, 3, 4	2

The foreman starts by assigning B to task 1, C to task 3 and D to task 2, but then realises there is a problem.

Draw a bipartite graph to model the situation and, using the maximum matching algorithm with the foreman's initial allocation as a starting point, find a complete matching of workers to tasks.

3 Aftab, Barry, Courtney and Dennis are to partner Edith, Fatima, Gladys and Hazel in a mixed doubles badminton tournament. Unfortunately, they are all very choosy with whom they play. The table shows those pairings which will not cause problems.

	E	F	G	H
A	✓			✓
B	✓		✓	
C	✓			
D		✓	✓	

Aftab suggests: 'I'll play with Edith. Barry can play with Gladys and Dennis with Fatima'. However, this leaves Courtney and Hazel, who don't get on.

Model the problem as a bipartite graph and, using the maximum matching algorithm with Aftab's suggestion as a starting point, find a complete matching.

4 Amy, Beulah, Caitlin and Daisy form a rowing team. There are four positions in the boat and the girls have the following preferences: Amy likes position 3 or 4; Beulah likes 1, 2 or 4; Caitlin likes 2 or 3; Daisy likes 1 or 2.

a) Show this information as a bipartite graph.

b) Show on your graph the partial matching, (Amy, 3), (Beulah, 1), (Daisy, 2).

c) Show that there are two possible alternating paths, starting from Caitlin. Hence, find the two possible complete matchings for the graph.

D1

5 A householder has a number of small building jobs that need doing, and gets quotes from four firms. The firms are not equipped to do all the jobs, and their quotes are shown in the table.

The jobs need to be done simultaneously, so it is necessary to give one job to each firm.

	Job 1	Job 2	Job 3	Job 4
Firm A	£200	–	£340	–
Firm B	£210	£110	–	–
Firm C	£180	–	–	£390
Firm D	–	£150	£320	£350

a) Model the situation as a bipartite graph, and hence find all possible ways in which the jobs could be allocated.

b) Hence, decide which is the cheapest plan for the householder.

Summary

You should know how to …

1 Model the relationship between two sets of items as a **bipartite graph**. The vertices of such a graph belong to two sets, and every edge connects a member of one set to a member of another set.

2 Find a **matching** for a bipartite graph, which pairs up vertices in one set with vertices in the other. The matching consists of a subset of the edges of the graph with the property that no two edges share a common vertex.

3 Recognise an **alternating path**, which joins an unmatched vertex in the left-hand set to an unmatched vertex in the right-hand set by edges which are alternately in and not in the set, M, of matched edges.

4 Improve a given partial matching using the **maximal matching algorithm**, which is as follows:

   Step 1   Start with a matching, M. (This could just be one edge, but it is usual to start with as good a matching as can quickly be found.)

   Step 2   Search for an alternating path. If none exists, go to Step 4.

   Step 3   Form a new matching by changing the status of the edges in the alternating path, and go to Step 2.

   Step 4   Stop. The current matching is maximal.

---

5 Find an alternating path. This can be found by inspection, perhaps aided by drawing a tree diagram, or by using one of the following algorithms.

Method 1

   Step 1   Choose an unmatched vertex in the left-hand set. Label it *.

   Step 2   For each most recently labelled left-hand vertex, identify all unlabelled right-hand vertices to which it is connected. If no such vertex exists, go to Step 8.

   Step 3   Label these with the name of the connected left-hand vertex.

   Step 4   If any of the vertices labelled in Step 3 is unmatched then you have a breakthrough. Go to Step 9.

   Step 5   For each most recently labelled right-hand vertex, identify any unlabelled left-hand vertex to which it is connected by an edge in the matched set, M. If no such vertex exists, go to Step 8.

   Step 6   Label these with the name of the connected right-hand vertex.

   Step 7   Go to Step 2.

   Step 8   If there are unlabelled unmatched vertices in the left-hand set, go to Step 1, or else stop – the matching is maximal.

   Step 9   Identify the alternating path by tracing back through the labels.

Method 2

Turn the graph into a directed network. Put arrows from right to left on all the edges in the matched set M, and from left to right on the edges not in M. Give each edge a weight of 1. The algorithm is then as follows:

   Step 1   Choose the unmatched vertex in the left-hand set and give it the permanent label 0.

   Step 2   Apply Dijkstra's algorithm until either an unmatched right vertex is reached (breakthrough) or no further labelling is possible.

   Step 3   If an unmatched right vertex is reached, trace the path back and stop – you have found an alternating path.

   Step 4   If any unmatched left vertices remain, go to Step 1. Otherwise stop – the matching is maximal.

Revision exercise 6

1 A group of five pupils have to colour in some pictures. There are five coloured crayons available, Orange (O), Red (R), Yellow (Y), Green (G) and Purple (P). The five pupils have each told their teacher their first and second choice of coloured crayons.

Pupil	First choice	Second choice
Alison (A)	Orange	Yellow
Brian (B)	Orange	Red
Carly (C)	Yellow	Purple
Danny (D)	Red	Purple
Emma (E)	Purple	Green

a) Show this information on a bipartite graph.

b) Initially, the teacher gives pupils A, C, D and E their first choice of crayons.
Demonstrate, by using an algorithm from this initial matching, how the teacher can give each pupil either their first or second choice of coloured crayons.

(AQA, 2001)

2 a) A group of five students is applying to five different universities. The students wish to visit the universities on 14 October but their teacher insists that no more than one student be allowed to visit the same university on that day. They list the two universities that they would like to visit.

Pupil	First choice	Second choice
Andrew (A)	Cambridge	Leeds
Joanne (J)	Cambridge	Durham
Rick (R)	Leeds	Bristol
Sarah (S)	Durham	Bristol
Tom (T)	Bristol	Oxford

i)   Draw a bipartite graph linking the students to their chosen two universities.

ii) Initially the teacher gives Andrew, Rick, Sarah and Tom their first choices of university.
Demonstrate, by using an algorithm from this initial matching, how the teacher can allocate each pupil to attend either their first or second choice of university.

(AQA, 2002)

D1

3 Six children are going to eat some fruit pastilles. There are six pastilles: blackcurrant (B), orange (O), plum (P), raspberry (R), strawberry (S) and tango (T). The six children will eat only certain flavours.

Name	Flavours
Alison (A)	B
Chris (C)	O, P, S
Derek (D)	B, P, R, S
Eddie (E)	O
Freda (F)	P, T
Gemma (G)	O, R, S

a) Show this information on a bipartite graph.

b) Initially, Chris chooses orange, Derek chooses blackcurrant, Freda chooses plum and Gemma chooses a strawberry pastille.
Demonstrate, by using an alternating path from this initial matching, how each child can get a pastille that they will eat.

(AQA, 2002)

4 a) Draw a bipartite graph representing the following adjacency matrix.

	1	2	3	4	5
A	1	0	1	0	0
B	0	0	1	1	0
C	0	1	0	1	0
D	0	0	1	1	0
E	1	1	0	0	1

b) Given that initially A is matched to 3, B is matched to 4 and E is matched to 1, use the maximum matching algorithm, from this initial matching, to find a complete matching. List your complete matching.

(AQA, 2003)

D1

7 Linear programming

This chapter will show you how to

◆ Express a problem as a linear programming formulation
◆ Illustrate graphically the range of choices which are consistent with the constraints of the problem
◆ Define the term 'objective function' and illustrate it on the graph
◆ Determine which combination of choices optimises the objective function

7.1 What is linear programming?

D1

The term linear programming refers to a set of mathematical techniques which were developed in the 1940s to assist the process of planning in a wide variety of industrial and economic situations. The aim is to choose the best combination of a number of quantities (variables) to optimise the outcome – typically, to maximise profit or minimise cost.

In a real-life context, there are likely to be many variables to be considered, but in this course you will not be required to deal with more than three.

Here is an example of a simple linear programming situation.

Example 1

Cloggs Breakfast Cereals make two types of muesli, Standard and De Luxe. A kilogram of Standard contains 800 grams of oat mix and 200 grams of fruit mix, while a kilogram of De Luxe has 600 grams of oat mix and 400 grams of fruit mix. A kilogram of Standard makes 60p profit, while a kilogram of De Luxe makes 80p.

The production planner knows that there are 3000 kg of oat mix and 1000 kg of fruit mix in stock. Past experience indicates that they will be able to sell at most 3500 kg of Standard and 2000 kg of De Luxe. The problem is to decide how much of each type to make.

· ·

These data can be summarised in a table.

Muesli type	Oat mix (kg)	Fruit mix (kg)	Max sales (kg)	Profit (£ per kg)
1 kg Standard	0.8	0.2	3500	0.60
1 kg De Luxe	0.6	0.4	2000	0.80
Availability	3000	1000		

The quantities to be decided are the amounts to produce of Standard and of De Luxe muesli. Let them be x kg and y kg, respectively. The variables x and y are called the **decision variables** (or **control variables**).

There are limitations on the possible values of x and y.

The known upper limits on sales tell you that:

$x \leqslant 3500$ and $y \leqslant 2000$

The amount of oat mix used will be $0.8x$ kg for Standard and $0.6y$ kg for De Luxe muesli. The stock of oat mix means that:

$0.8x + 0.6y \leqslant 3000$

There are 3000 kg of oat mix.

which simplifies to give:

$4x + 3y \leqslant 15\,000$

The amount of fruit mix used will be $0.2x$ kg for Standard and $0.4y$ kg for De Luxe muesli. The stock of fruit mix means that:

$0.2x + 0.4y \leqslant 1000$

There are 1000 kg of fruit mix.

which simplifies to give:

$x + 2y \leqslant 5000$

These limitations are called the **constraints** of the problem. There are two more, trivial, constraints: namely, $x \geqslant 0, y \geqslant 0$.

The profit on x kg of Standard and y kg of De Luxe muesli is:

$P = 0.6x + 0.8y$

This is called the **objective function**. The aim is to maximise the value of P.

The problem can now formally be stated as a **linear programming (LP) formulation**:

Maximise $\qquad\qquad\qquad P = 0.6x + 0.8y$

subject to the constraints $\quad x \leqslant 3500$
$\qquad\qquad\qquad\qquad\quad y \leqslant 2000$
$\qquad\qquad\qquad\qquad\quad 4x + 3y \leqslant 15\,000$
$\qquad\qquad\qquad\qquad\quad x + 2y \leqslant 5000$
$\qquad\qquad\qquad\qquad\quad x \geqslant 0, y \geqslant 0$

You will explore the solution to this problem on pages 106 to 108, but for the moment here are some other examples of formulating linear programming problems.

Example 2

A waste-paper merchant has two processing plants. Plant A can process 4 tonnes of waste paper and 1 tonne of cardboard per hour, while Plant B can process 5 tonnes of waste paper and 2 tonnes of cardboard per hour. It costs £400 per hour to run plant A and £600 per hour for plant B. The merchant has an agreement with the union that no consignment should be shared out in such a way that one plant gets more than twice the running time of the other.

A consignment arrives, consisting of 100 tonnes of waste paper and 35 tonnes of cardboard. How should this be shared between the two plants in order to minimise the processing costs?

Most of these data can be summarised in a table:

Plant type	Waste paper (tonnes)	Cardboard (tonnes)	Cost (£)
Plant A (1 hour)	4	1	400
Plant B (1 hour)	5	2	600
Availability	100	35	

First, **state the decision variables**. These are the running times for the two plants:

Run A for x hours and B for y hours

Next, **set the constraints**:

Because of the union agreement, $x \leqslant 2y$ and $y \leqslant 2x$

The two plants must between them be able to process at least 100 tonnes of paper in the allotted time, giving:

$4x + 5y \geqslant 100$

Similarly, the need to process at least 35 tonnes of cardboard means that:

$x + 2y \geqslant 35$

There are also the usual non-negativity constraints:

$x \geqslant 0$ and $y \geqslant 0$.

Lastly, **state the objective function**. In this case, the objective is to minimise the cost. So the objective function is:

$C = 400x + 600y$

Now, formally state the problem:

Minimise $C = 400x + 600y$

subject to the constraints $x \leqslant 2y$
$y \leqslant 2x$
$4x + 5y \geqslant 100$
$x + 2y \geqslant 35$
$x \geqslant 0, y \geqslant 0$

> $x \leqslant 2y$ means that A's running time is not more than twice B's running time.

D1

Example 3

A furniture company sells three types of dining chair. They obtain these by contracting to buy a whole week's production from a manufacturer working to their specifications. All of the chairs pass through three workshops for cutting, assembly and finishing. Each workshop runs for a working week of 40 hours. The times in the workshops for each type of chair and the profit gained from each are shown in the table. How many of each should be made to maximise the weekly profit?

Chair type	Cutting (hours)	Assembly (hours)	Finishing (hours)	Profit (£)
Type A	0.5	1	0.75	60
Type B	0.75	1.25	0.75	80
Type C	0.5	0.75	0.5	50

First, **state the decision variables**. These are numbers of each type of chair.

Let x be the number of Type A, y the number of Type B and z the number of Type C.

Next, **set the constraints**.

The time in the cutting shop means that:

$0.5x + 0.75y + 0.5z \leqslant 40$
which simplifies to give: $2x + 3y + 2z \leqslant 160$

The time in the assembly shop means that:

$x + 1.25y + 0.75z \leqslant 40$
which simplifies to give: $4x + 5y + 3z \leqslant 160$

The time in the finishing shop means that:

$0.75x + 0.75y + 0.5z \leqslant 40$
which simplifies to give: $3x + 3y + 2z \leqslant 160$

There are the usual non-negativity constraints: $x \geqslant 0, y \geqslant 0$ and $z \geqslant 0$.

However, in this example, there is also the constraint that you cannot make use of fractions of a chair, so x, y and z must be integers.

Last, **state the objective function**. In this case, the objective is to maximise the profit, so the objective function is:

$P = 60x + 80y + 50z$

Now, formally state the problem:

Maximise $\qquad\qquad P = 60x + 80y + 50z$

subject to the constraints $2x + 3y + 2z \leqslant 160$
$\qquad\qquad\qquad\qquad\quad 4x + 5y + 3z \leqslant 160$
$\qquad\qquad\qquad\qquad\quad 3x + 3y + 2z \leqslant 160$
$\qquad\qquad\qquad\qquad\quad x \geqslant 0, y \geqslant 0, z \geqslant 0$
$\qquad\qquad\qquad\qquad\quad x, y$ and z are integers

D1

> There are three constraints, one for each workshop.

Finally, a blending problem will be examined.

Example 4

A company supplying vegetable oil buys from two sources, A and B. The oils supplied are blends of olive oil, sunflower oil and other vegetable oils. The table shows the proportions, price and minimum weekly order of these.

Source	Olive oil	Sunflower oil	Other	Cost (p per litre)	Minimum order (litres)
A	50%	10%	40%	25	35 000
B	20%	60%	20%	20	50 000

The company is to make a blend with at least 30% olive oil and at least 40% sunflower oil. They want to minimise the cost.

· ·

First, **state the decision variables**. These are the amounts of the two types to use:

Use x litres of A and y litres of B.

Next, **set the constraints**.

The requirement for at least 30% olive oil gives:

$$\frac{0.5x + 0.2y}{x + y} \geq 0.3$$

This simplifies to give:

$$y \leq 2x$$

The requirement for at least 40% sunflower oil gives:

$$\frac{0.1x + 0.6y}{x + y} \geq 0.4$$

which simplifies to give:

$$2y \geq 3x$$

The minimum order requirements give:

$$x \geq 35\,000 \quad \text{and} \quad y \geq 50\,000$$

Last, **state the objective function**. In this case, the objective is to minimise the cost. So, the objective function is $C = 25x + 20y$.

Now, formally state the problem:

Minimise $\qquad\qquad C = 25x + 20y$

subject to the constraints $\quad y \leq 2x$
$\qquad\qquad\qquad\qquad\quad 2y \geq 3x$
$\qquad\qquad\qquad\qquad\quad x \geq 35\,000$
$\qquad\qquad\qquad\qquad\quad y \geq 50\,000$

Simplify the inequality:
$0.5x + 0.2y \geq 0.3x + 0.3y$
$5x + 2y \geq 3x + 3y$
$2x \geq y$
$y \leq 2x$

Simplify the inequality:
$0.1x + 0.6y \geq 0.4x + 0.4y$
$x + 6y \geq 4x + 4y$
$2y \geq 3x$

Exercise 7A

For each of the following situations state the problem as a linear programming formulation.

1 Roger Teeth Ltd make fruit drinks of two types, consisting of fruit juice, sugar syrup and water. A litre of their Econofruit drink contains 0.2 litres of juice, 0.5 litres of syrup and the rest water. A litre of Healthifruit contains 0.4 litres of juice, 0.3 litres of syrup and the rest water. The profit per litre is 30p for Econofruit and 40p for Healthifruit. They have 20 000 litres of juice and 30 000 litres of syrup in stock (and, of course, unlimited water).
They wish to maximise the profit.

2 A club, with 80 members, is organising a trip. The intention is to hire vehicles they can drive themselves and travel in convoy. Only eight of the members are prepared to drive. A car, which can carry five people including the driver, costs £20 per day to hire. A minibus, which can carry 12 people including the driver, costs £60 per day to hire.
The club wants to minimise the hire costs.

3 A farmer has 75 hectares of land on which to grow a mixture of wheat and potatoes. Each hectare of wheat requires 30 man-hours of labour and 700 kg of fertiliser, while each hectare of potatoes needs 50 man-hours of labour and 400 kg of fertiliser. There are 2800 man-hours of labour and 40 tonnes of fertiliser available. Wheat realises a profit of £80 per hectare, and potatoes £100 per hectare.
The farmer's aim is to maximise his profit.

4 A fruiterer is making up baskets of fruit, which are advertised as containing a mixture of oranges, apples and pears with at least 30 fruit in all. There must be at least as many apples as oranges, and at least twice as many apples as pears. Oranges cost 20p each, apples 12p each and pears 15p each.
The fruiterer's aim is to minimise her cost.

5 Whisky mac is a mixture of whisky and ginger wine. Whisky is 40% alcohol and costs £12 per litre. Ginger wine is 12% alcohol and costs £5 per litre.
A barkeeper wishes to minimise the cost of making a whisky mac, which must be at least 100 ml of liquid, at least 20% alcohol and contain at most 30 ml of alcohol.

7.2 Solving linear programming problems

Problems with two decision variables, and those which can be reduced to two variables, will now be considered. Such problems can be solved graphically.

Take the problem of optimising the production of muesli, which was detailed in Example 1 (pages 100 to 101). When stated as a linear programming formulation, it gives:

Maximise $P = 0.6x + 0.8y$

Subject to the constraints $x \leqslant 3500$
$y \leqslant 2000$
$4x + 3y \leqslant 15\,000$
$x + 2y \leqslant 5000$
$x \geqslant 0, y \geqslant 0$

D1

where x kg and y kg are the amounts of Standard and De Luxe muesli produced.

Illustrating the constraints

You can illustrate the range of choice available to the production planner with a graph. First, consider the constraints $x \leqslant 3500$ and $y \leqslant 2000$. These can best be illustrated by drawing the lines $x = 3500$ and $y = 2000$ and then shading out the regions which are not required.

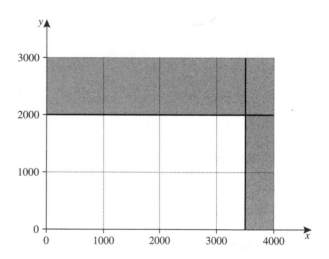

The constraint $4x + 3y \leqslant 15\,000$ can now be added. Draw the line $4x + 3y = 15\,000$, and then shade the region which does not satisfy the inequality. (The usual convention with inequalities is to draw the boundary line dashed for $<$ or $>$, and continuous for $\leqslant$ and $\geqslant$.)

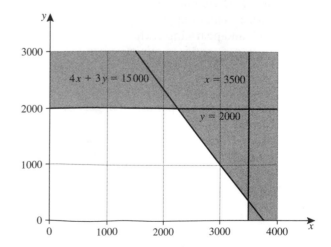

Now, add the constraint $x + 2y \leqslant 5000$ by drawing
the line $x + 2y = 5000$ and then shading the
unwanted region.

You can also show the trivial constraints $x \geqslant 0$
and $y \geqslant 0$ by shading the negative values, as in the
diagram on the right, but this is usually assumed.

Any allowable combination of x and y must lie in
the unshaded region of the graph (including its
boundary lines). This region is called the **feasible
region**.

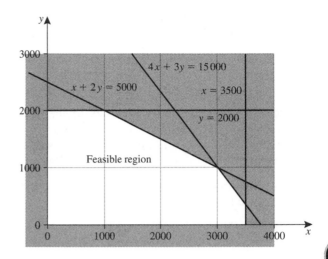

Illustrating the objective function

Any pair of values of x and y corresponding to a
point in the feasible region represents a possible
production plan, although clearly some points will
be better than others. For example, the point
$x = 1000, y = 1000$ lies in the feasible region, but it
would not be the best production plan, as it would
leave a lot of unused oat mix and fruit mix.

If you opted for the (unsatisfactory) production
plan of $x = 1000$ and $y = 1000$, you would have:

$$P = 0.6 \times 1000 + 0.8 \times 1000 = 1400$$

So, this production plan would realise £1400
profit.

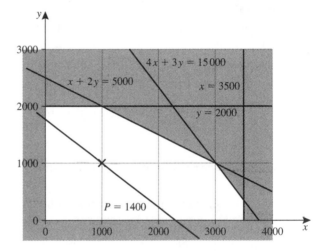

There are other production plans giving $P = 1400$
for example, $x = 0$ and $y = 1750$. These plans all lie
on the line $0.6\,x + 0.8\,y = 1400$, which can be
drawn on your graph.

Similarly, you could draw a line $0.6x + 0.8y = 1800$,
showing all the production plans which would give
a profit of £1800.

These are two possible positions for the **objective
line**.

You can see that increasing profit corresponds to
moving the objective line away from the origin,
always keeping the same gradient. As long as at
least one point in the feasible region lies on this
line, there is a production plan which will give that
amount of profit. The problem, therefore, is to find
the most extreme position of the objective line
which still includes a point in the feasible region.

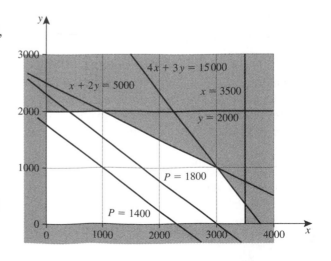

D1

It is clear that, as the objective line is moved away from the origin, its last contact with the feasible region will be at one of its vertices.

In simple cases it is possible to decide which vertex this is by drawing one position of the objective line and then sliding a ruler parallel to this.

It is clear from the graph that, in this example, the vertex formed by the intersection of the lines $x + 2y = 5000$ and $4x + 3y = 15\,000$ will give the maximum value of P. By solving these simultaneous equations, you can confirm that this is the point $(3000, 1000)$.

This corresponds to a production plan for 3000 kg of Standard muesli and 1000 kg of De Luxe muesli. The profit is then

$$0.6 \times 3000 + 0.8 \times 1000 = £2600$$

> If the objective line is parallel to one of the boundaries, then all the points on that boundary will give the optimal value for the objective function.

> Or you can substitute the coordinates of the vertices into the objective function to establish which one gives the best value.

D1

Example 5

Find a graphical solution for the following linear programming problem.

Minimise	$C = 400x + 600y$
subject to the constraints	$x \leqslant 2y$
	$y \leqslant 2x$
	$4x + 5y \geqslant 100$
	$x + 2y \geqslant 35$
	$x \geqslant 0, y \geqslant 0$

(This is the waste-paper problem introduced in Example 2, page 102.)

· ·

Draw a graph showing these constraints, with the unwanted regions shaded, as shown.

Also draw one possible position of the objective line. For example, if $x = 20$, $y = 20$, a point which is clearly in the feasible region, you will get $C = 20\,000$. So, draw the line

$$400x + 600y = 20\,000$$

as shown.

Moving the objective line towards the origin parallel to itself will reduce the value of C. However, in this case it is not easy to see from the graph which of the vertices will give the minimum value of C. Therefore, you use simultaneous equations to find the coordinates of the vertices.

Solving $y = 2x$ and $4x + 5y = 100$ gives $x = 7\frac{1}{7}$, $y = 14\frac{2}{7}$, which yield:

$$C = 400 \times 7\frac{1}{7} + 600 \times 14\frac{2}{7} = 11\,428\frac{4}{7}$$

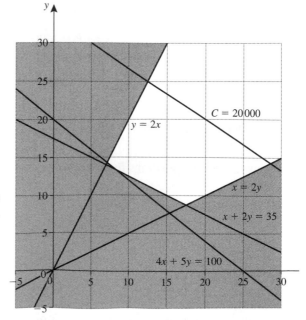

Solving $4x + 5y = 100$ and $x + 2y = 35$ gives $x = 8\frac{1}{3}, y = 13\frac{1}{3}$, which yield:

$$C = 400 \times 8\frac{1}{3} + 600 \times 13\frac{1}{3} = 11\,333\frac{1}{3}$$

Solving $x + 2y = 35$ and $x = 2y$ gives $x = 17\frac{1}{2}, y = 8\frac{3}{4}$, which yield:

$$C = 400 \times 17\frac{1}{2} + 600 \times 8\frac{3}{4} = 12\,250$$

You can now see that the minimum value of C is $11\,333\frac{1}{3}$, which occurs when $x = 8\frac{1}{3}, y = 13\frac{1}{3}$. The solution to the waste-paper merchant's problem is, therefore, to run plant A for $8\frac{1}{3}$ hours, processing $33\frac{1}{3}$ tonnes of paper and $8\frac{1}{3}$ tonnes of cardboard; and to run plant B for $13\frac{1}{3}$ hours, processing $66\frac{2}{3}$ tonnes of paper and $26\frac{2}{3}$ tonnes of cardboard.

D1

Example 6

Find a graphical solution of the following linear programming problem.

Maximise $\qquad\qquad P = 42x + 20y$

subject to the constraints $\quad x + 2y \leqslant 14$

$\qquad\qquad\qquad\qquad\qquad 3x + y \leqslant 20$

$\qquad\qquad\qquad\qquad\qquad x \geqslant 0, y \geqslant 0$

$\qquad\qquad\qquad\qquad\qquad x$ and y are integers

> The equation of the objective line is $P = 42x + 20y$. The point $x = 5, y = 0$ is in the feasible reason and gives a value of $P = 210$.

Draw a graph showing the constraints, shading the unwanted region. As you are constrained to integer values of x and y, only the points indicate by dots in the feasible region can be valid solutions.

Draw one possible position of the objective line, corresponding, in this case, to $P = 210$.

You can see from the graph that, as the objective line is moved away from the origin, increasing the value of P, its final contact with the feasible region will be at the intersection of the lines $x + 2y = 14$ and $3x + y = 20$. However, these lines intersect at (5.2, 4.4), which is not an integer point.

It is therefore necessary to test all likely integer points near to (5.2, 4.4) to find which gives the maximum value of P.

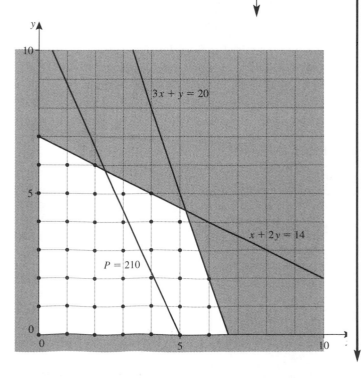

The obvious points to test are $(4, 5)$, $(5, 4)$ and $(6, 2)$.

$x = 4$ and $y = 5$ give $P = 42 \times 4 + 20 \times 5 = 268$
$x = 5$ and $y = 4$ give $P = 42 \times 5 + 20 \times 4 = 290$
$x = 6$ and $y = 2$ give $P = 42 \times 6 + 20 \times 2 = 292$

Hence, the optimum value of P is 292, obtained when $x = 6$ and $y = 2$.

Problems with three variables

Generally, when a problem has three variables, it cannot be solved graphically. You may be asked in the examination to formulate a problem with three variables, but its solution lies outside the syllabus.

However, sometimes in a three-variable problem, the variables are related in such a way that they can be reduced to two. Hence, the problem can be solved graphically.

D1

Example 7

A dog food manufacturer makes three types of chew, each 10 g in weight, which are made from different proportions of two basic ingredients. The table shows this, together with the amount of the two ingredients in stock, and the costs for the three types of chew.

Chew type	Ingredient 1	Ingredient 2	Cost (p per chew)
Type A	8 g	2 g	1.8
Type B	6 g	4 g	1.6
Type C	5 g	5 g	1.5
Availability	800 kg	400 kg	

The manufacturer wants to make 1600 packets of mixed chews. Each must contain 60 chews, and there must be no more than 30 of each type of chew in a packet. How many of each type should be included to minimise the cost?

· ·

Decision variables
Each packet contains x of Chew A, y of Chew B, z of Chew C.

Constraints
Ingredient 1 availability means:
$1600(8x + 6y + 5z) \leq 800\,000$
which simplifies to:
$8x + 6y + 5z \leq 500$
Ingredient 2 availability means:
$1600(2x + 4y + 5z) \leq 400\,000$
which simplifies to:
$2x + 4y + 5z \leq 250$
No more than 30 of each type means: $x \leq 30$, $y \leq 30$, $z \leq 30$.

> $1600(8x + 6y + 5z)$ is the total weight of Ingredient 1 in 1600 packets.

> $1600(2x + 4y + 5z)$ is the total weight of Ingredient 2 in 1600 packets.

60 chews per packet means: $x + y + z = 60$.
Non–negativity constraints: $x \geq 0$, $y \geq 0$, $z \geq 0$.
Only whole numbers are possible. Hence, x, y, z are integers.

Objective function
The cost is given by:

$C = 1.8x + 1.6y + 1.5z$

Hence, in terms of three variables, the problem can be stated as:

Minimise $\qquad\qquad\qquad C = 1.8x + 1.6y + 1.5z$

subject to the constraints $\quad 8x + 6y + 5z \leq 500$
$\qquad\qquad\qquad\qquad\qquad 2x + 4y + 5z \leq 250$
$\qquad\qquad\qquad\qquad\qquad x \leq 30,\ y \leq 30,\ z \leq 30$
$\qquad\qquad\qquad\qquad\qquad x + y + z = 60$
$\qquad\qquad\qquad\qquad\qquad x \geq 0,\ y \geq 0,\ z \geq 0$
$\qquad\qquad\qquad\qquad\qquad x, y, z$ are integers

However, because $x + y + z = 60$, you have $z = 60 - x - y$.
Substituting this into the objective function, together with the
constraints, gives a problem with two variables, as follows.

$\begin{aligned} C &= 1.8x + 1.6y + 1.5z \\ &= 1.8x + 1.6y + 1.5(60 - x - y) \\ &= 90 + 0.3x + 0.1y \end{aligned}$

$8x + 6y + 5z \leq 500$ gives $8x + 6y + 5(60 - x - y) \leq 500$
$\qquad\qquad\qquad\qquad\qquad\qquad\quad 3x + y \leq 200$

$2x + 4y + 5z \leq 250$ gives $2x + 4y + 5(60 - x - y) \leq 250$
$\qquad\qquad\qquad\qquad\qquad\qquad\quad 3x + y \geq 50$

$z \leq 30$ gives $60 - x - y \leq 30$
$\qquad\qquad\quad \Rightarrow\ x + y \geq 30$

$z \geq 0$ gives $60 - x - y \geq 0$
$\qquad\qquad\quad \Rightarrow\ x + y \leq 60$

Hence, in terms of two variables, the
problem can be stated and illustrated as:

Minimise $\qquad\qquad C = 90 + 0.3x + 0.1y$

subject to the constraints $\quad 3x + y \leq 200$
$\qquad\qquad\qquad\qquad\qquad 3x + y \geq 50$
$\qquad\qquad\qquad\qquad\qquad x \leq 30, y \leq 30$
$\qquad\qquad\qquad\qquad\qquad x + y \geq 30$
$\qquad\qquad\qquad\qquad\qquad x + y \leq 60$
$\qquad\qquad\qquad\qquad\qquad x \geq 0, y \geq 0$
$\qquad\qquad\qquad\qquad\qquad x, y$ are integers

You can see from the graph that the constraints
$3x + y \leq 200$ and $x + y \leq 60$ are unnecessary.

Redraw the graph without these, showing one
possible position of the objective line,
corresponding, in this case, to $C = 98$.

The point (20, 20) is in the
feasible region, and gives a
value of
$C = 90 + 0.3 \times 20 + 0.1 \times 20$
$\quad = 98$

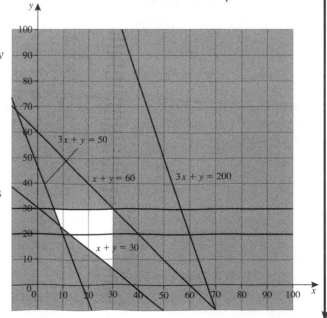

D1

It is clear that the objective line is parallel to the boundary $3x + y = 50$, so any integer point on this boundary will minimise the objective function. The possible points are (7, 29), (8, 26), (9, 23) and (10, 20). Therefore, the possible contents of the packets of chews are as given in the table below.

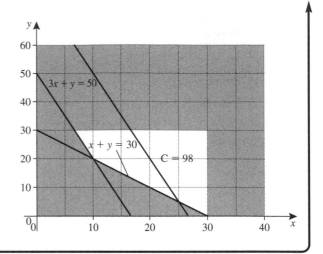

Chew A	Chew B	Chew C
7	29	24
8	26	26
9	23	28
10	20	30

D1

Exercise 7B

1 For each of the following linear programming problems, draw a graph showing the feasible region, and one possible position of the objective line. Hence, find the optimal value of the objective function and the corresponding values of x and y.

a) Maximise $P = 3x + 2y$

    subject to $3x + 4y \leqslant 120$
               $3x + y \leqslant 75$
               $x \geqslant 10, y \geqslant 5$

b) Maximise $P = 4x + y$

    subject to $x + y \leqslant 10$
               $2x + y \leqslant 16$
               $x \geqslant 0, y \geqslant 0$

c) Minimise $C = 5x + 4y$

    subject to $15x + 8y \geqslant 90$
                $y \geqslant x$
                $y \leqslant 2x$
                x, y are integers

d) Maximise $R = 2x + y$

    subject to $x + y \leqslant 14$
                $x + 2y \leqslant 20$
                $2x + 3y \leqslant 32$
                $x \geqslant 0, y \geqslant 0$

2 In Example 4 on page 104, you explored the problem of a company blending oil from two sources, A and B. The linear programming formulation was:

    Minimise $C = 25x + 20y$

    subject to the constraints $y \leqslant 2x$
                                   $2y \geqslant 3x$
                                   $x \geqslant 35\,000$
                                   $y \geqslant 50\,000$

Use graphical methods to decide how much oil the company should buy from each source.

3 In Exercise 7A, Question **1** (page 105), you expressed, as a linear programming (LP) formulation, the problem of deciding the most profitable quantities of Econofruit and Healthifruit drinks to make. Draw a graph showing the feasible region and hence find the optimum combination.

4 In Exercise 7A, Question **2**, you expressed, as an LP formulation, the problem of deciding the cheapest combination of minibuses and cars to hire. Draw a graph showing the feasible region and hence find the optimum combination.

> **Remember:** LP means 'linear programming'.

5 In Exercise 7A, Question **3** , you expressed, as an LP formulation, the problem of maximising profit from growing wheat and potatoes. Use graphical methods to obtain the optimum planting strategy.

D1

6 A building firm has a plot of land, the area of which is 6000 m². The intention is to build a mixture of houses and bungalows. A house occupies 210 m² and a bungalow occupies 270 m². Local planning regulations limit the number of dwellings that can be built to 25, and insist that there must be no more than 15 of either type. Each house will realise £20 000 profit, and each bungalow £25 000. Formulate this situation as a linear programming problem and find graphically the best combination of dwellings.

7 It is planned to take 50 people on a trip. The party comprises x senior staff, y trainees and z children. There must be at least one adult to every two children, and at least one senior staff member to every trainee. There must be at least five trainees and at least ten senior staff. The cost of the trip is £20 for each senior staff member, £15 for each trainee and £12 for each child. It is required to minimise the total cost.

a) Express this as an LP formulation in x, y and z.

b) Using the fact that $x + y + z = 50$, show that your formulation can be expressed as:

$$\text{Minimise} \quad C = 8x + 3y + 600$$

$$\text{subject to} \quad 3x + 3y \geqslant 50$$
$$x \geqslant y$$
$$x \geqslant 10, y \geqslant 5$$
$$x, y \text{ are integers}$$

c) Use graphical methods to find the optimum combination of people.

Summary

You should know how to ...

1 In a linear programming problem, choose the values of the decision (or control) variables in order to maximise or minimise the **objective function** (usually profit or cost).

2 Set out the **constraints** which limit the values of the decision variables.

3 Express a problem as a **linear programming (LP) formulation**.

4 Draw a graph to illustrate a linear programming problem.

5 Determine values of the objective function which correspond to different positions of the **objective line** and find the optimal solution.

D1

Revision exercise 7

1 A company is making two types of door, standard and luxury. Both types of door require the use of two different machines, A and B.

Both types of door require 90 minutes on machine A. A standard door requires 60 minutes on machine B but a luxury door requires 120 minutes on this machine.

During any one week, machine A can be used for a maximum of 20 hours and machine B can be used for a maximum of 25 hours.

The company makes a profit of £10 on each standard door and £12 on each luxury door. In a week, the company makes x standard door and y luxury doors.

a) Show that the above information can be modelled by the following inequalities.

$$x \geqslant 0 \quad y \geqslant 0 \quad 3x + 3y \leqslant 40 \quad x + 2y \leqslant 25$$

b) i) Draw a suitable diagram to represent the problem graphically, indicating the feasible region.

   ii) Draw an objective line that will represent the company's profit for the week, £P. Hence, indicate the vertex, V, that could correspond to the maximum value of P.

   iii) State why this maximum value of P cannot be achieved.

   iv) Find values of x and y that will maximise the company's profit and calculate this profit.

(AQA, 1999)

2 A factory manufactures three items: screws, nuts and bolts.

The items are first produced on a machine which is available for
4 hours per day. The machine takes 6 minutes to produce a screw,
4 minutes to produce a nut and 2 minutes to produce a bolt. The
items are then cleaned on a machine that is available for 55 minutes
per day. Each screw takes 30 seconds to clean, each nut takes 40
seconds to clean and each bolt takes 60 seconds to clean.

a) An apprentice at the factory produces x screws, y nuts and z bolts
in a day. Find and simplify **two** inequalities, each involving x, y
and z, that model the conditions given above.

b) As a further requirement, the apprentice must produce the same
number of bolts as nuts each day.
  i)   Show that, with this additional constraint, the two
       inequalities found in part a) become:
$$x + y \leqslant 40 \quad \text{and} \quad 3x + 10y \leqslant 330$$
  ii) Draw a suitable diagram to represent this problem
       graphically, indicating the feasible region.
  iii) The apprentice has to make the largest possible total number
       of items each day. Draw an objective line that will represent
       his total daily output, T. Hence, indicate the vertex that will
       correspond to the maximum value of T.
  iv) Calculate the maximum value of T.
  v)   On a particular day, the cleaning machine is available for
       only 48 minutes. Find the maximum number of items that
       the apprentice can make during this particular day.

(AQA, 1999)

D1

3 The graph shows the feasible region of a linear
programming problem.

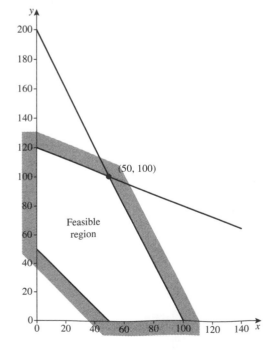

a) On this feasible region find:
   i)   the maximum value of the function $2x + 3y$
   ii) the minimum value of the function $4x + y$

b) Find the **five** inequalities that define the feasible region.   (*AQA, 2002*)

4 A tropical fish collector wishes to fill an aquarium with two types of fish, X-ray fish and Yellowtails. He requires that:

   There are at least two X-ray fish.
   There are at least as many Yellowtails as X-ray fish.
   There are at least 6 fish and at most 10 altogether.

a) Let x be the number of X-ray fish and y the number of Yellowtails. Write down inequalities in x and y which represent the above conditions.

b) On graph paper illustrate the region of those x and y which satisfy the inequalities in part a).

c) The cost of each X-ray fish is £2 and the cost of each Yellowtail is £4. Find the minimum and maximum total costs of a collection of fish satisfying the above conditions.

d) List all the possible total costs of a collection of fish satisfying the above conditions.   (*AQA, 2002*)

5 The Carryit Company has 1000 large packing cases to transport from Liverpool to Southampton; x of them by road, y of them by rail, and the rest by sea.

The company intends to transport at least 10% and at most 50% of the cases by sea.

Furthermore, the number transported by sea must be at most double the number transported by road.

a) Write down, in terms of x and y, the number of cases transported by sea. Hence show that x and y must satisfy the inequalities:

$$500 \leqslant x + y \leqslant 900 \quad \text{and} \quad 3x + y \geqslant 1000$$

b) Illustrate on graph paper the region of those (x, y) which satisfy $x \geqslant 0$ and $y \geqslant 0$ together with the inequalities in part a).

c) The cost of transporting the cases is:

>   £50 for each case by road;
>   £55 for each case by rail; and
>   £25 for each case by sea.

   i)   Show that the total cost £T of sending the 1000 cases is given by:

$$T = 25x + 30y + 25\,000.$$

   ii) Hence use part b) to find the minimum cost of transporting all the cases. State how many of the cases should be sent by each method in order to achieve that minimum.

   iii) In order to be more competitive the rail operator wants to reduce the cost of transporting each case to £R, where R is a whole number. The costs of transporting by road and sea will remain the same.

   The rail operator chooses R as large as possible so that, when minimising its transport costs, the Carryit Company will have to send most of the 1000 cases by rail. Calculate the value of R.   *(AQA, 2003)*

D1

6 The Bumper Hamper Company offers a selection of hampers containing tins of chicken, pots of caviar and jars of peaches.

   ✦ Each tin of chicken weighs 300 grams, each pot of caviar weighs 100 grams and each jar of peaches weighs 200 grams.
   ✦ Each hamper contains at most 11 items and weighs at most 2200 grams.
   ✦ The company makes £4 profit on each tin of chicken, £3 profit on each pot of caviar and £1 profit on each jar of peaches.

Let a hamper contain x tins of chicken, y pots of caviar and z jars of peaches.

a) Express each of the **two** restrictions on a hamper's contents as an inequality in x, y and z.

b) Write down the total profit £P on a hamper in terms of x, y and z.

c) The company decides to include just one jar of peaches in each hamper, but then to choose the amount of chicken and caviar in order to maximise its profits. Show that this is equivalent to maximising:

$$P = 4x + 3y + 1$$

subject to:

$$x \geqslant 0 \quad y \geqslant 0 \quad x + y \leqslant 10$$

and **one** other inequality, which you should state.

d) Use a graphical method to solve the linear programming problem in part c).

e) The company drops the restriction of having to include peaches in each hamper. Find the maximum profit it can then make on a hamper.   *(AQA, 2003)*

7 A strawberry grower decides to make jars of jam and jars of jelly for sale. The required ingredients for the two recipes are:

Jam recipe (makes 2 jars)	Jelly recipe (makes 10 jars)
0.5 kg strawberries	6 kg strawberries
1 kg sugar	7 kg sugar
1 lemon	2 lemons

The strawberry grower has 45 kg of strawberries, 60 kg of sugar and 40 lemons.

Assume that he makes x quantities of the jam recipe and y quantities of the jelly recipe.

a) Explain why $\frac{1}{2}x + 6y \leqslant 45$ and write down two other similar inequalities which x and y must satisfy.

b) Illustrate the region representing those (x, y) satisfying $x \geqslant 0, y \geqslant 0$ and the three inequalities from part a).

c) Identical jars are used for both the jam and the jelly. Find the maximum number of jars which the grower might need.

d) The grower intends to make an equal number of jars of jam and jelly. Find the maximum number of jars of each which he can make.

(AQA, 2002)

8 A market trader sells sets of pans and sets of crockery.

He wants the number of sets of crockery stocked to be at least twice the number of sets of pans stocked.

He can fit at most 28 sets in total on his stall.

Each set of pans weighs 1 kilogram, each set of crockery weighs 2 kilograms, and his stall can hold a maximum of 50 kilograms.

a) Introduce two variables and write each of the above conditions as an inequality in those variables.

b) Illustrate a feasible region of those points which are non-negative and satisfy the inequalities in part a).

c) Each set of pans makes him £2 profit and each set of crockery makes him £3 profit. How many sets of each should he stock in order to maximise his profit? Calculate this maximum profit.

d) The trader manages to buy his pans from a different wholesaler so that his profit on a set of pans goes up to £4. How does that affect the number of sets of pans and crockery which he should stock?

(AQA, 2001)

9 A factory can make three types of sofa: Chesterfields, Darleys and Edales. Customer demand and availability of the workforce mean that the following conditions apply.

At least 20 of each type must be made each week.

Each Chesterfield and each Darley take 20 man-hours to make and each Edale takes 30 man-hours. There are 2000 man-hours available each week.

At most, half the sofas made each week should be Edales.

Let c be the number of Chesterfields produced in one week, d the number of Darleys produced, and e the number of Edales produced.

a) Express the given conditions as inequalities in c, d and e.

b) Deduce from your inequalities in a) that:

$$2e \leqslant 2(c + d) \leqslant 200 - 3e$$

c) Hence, show that the maximum number of Edales which can be made each week is 40. In order to make 40 Edales each week, how many Chesterfields and how many Darleys should be made?

(AQA, 2003)

D1

8 Sorting

This chapter will show you how to

◆ Sort a list using the bubble sort, shuttle sort, Shell sort and Quicksort algorithms
◆ Compare the methods in terms of the number of operations required

8.1 Sorting algorithms

One of the most common data processing tasks is the sorting of an unordered list into numerical or alphabetical order. Although you can easily sort a short list by hand, using rather haphazard and intuitive methods, for long lists and for computer implementation, you need an algorithm.

> In computing terms, these amount to much the same thing, as text is stored in the form of number codes in a computer.

There are many sorting algorithms – a cursory search on the Internet will find 20 or more. These all involve comparing items in the list and moving or labelling them to arrive at the required order. They vary enormously in the speed with which they achieve their objective, and there is no single best method for all circumstances. Four of the simpler algorithms are described in detail in this chapter.

Bubble sort

This involves passing along the list a number of times, comparing and, if necessary swapping pairs of numbers. The effect is that after each pass the next largest number has moved to its rightful place.

> This gives the algorithm its name, after the rather fanciful notion that the numbers rise in the list like bubbles rising in a glass of lemonade.

The **bubble sort** algorithm for a list of numbers can be written as follows.

Work along the list (make a 'pass' through the list) a number of times according to these rules:

◆ **First pass**
Compare the first and second numbers. Swap them if they are in the wrong order.

Successively compare/swap the second and third numbers, the third and fourth numbers, and so on, to the end of the list.

◆ **Subsequent passes**
Repeat the above process, leaving out the final number from the previous pass.

◆ **Terminating**
The process finishes either when the remaining list has only one number or when a complete pass produces no swaps.

An example will show this in detail.

Example 1

Arrange the numbers 4, 8, 2, 6, 3, 5 in ascending order, using the bubble sort algorithm.

The shading shows the comparison being made at each stage, with the arrow indicating when a swap is necessary.

First pass

4	8	2	6	3	5
4	8 ⟷ 2		6	3	5
4	2	8 ⟷ 6		3	5
4	2	6	8 ⟷ 3		5
4	2	6	3	8 ⟷ 5	
4	2	6	3	5	8

> 4 < 8, so no swap necessary.
> 8 is greater than the other numbers in the list, so is swapped repeatedly.

Second pass

4 ⟷ 2		6	3	5	8
2	4	6	3	5	8
2	4	6 ⟷ 3		5	8
2	4	3	6 ⟷ 5		8
2	4	3	5	6	8

> 4 > 2, so swap.
> 4 < 6, no swap.
> 8 was the final number in the first pass, so no need to consider it again.

Third pass

2	4	3	5	6	8
2	4 ⟷ 3		5	6	8
2	3	4	5	6	8
2	3	4	5	6	8

> 6 and 8 do not need to be considered in the third pass.

Fourth pass

2	3	4	5	6	8
2	3	4	5	6	8
2	3	4	5	6	8

As there were no swaps on the fourth pass, you stop the process.

Efficiency

In any given situation, the different sorting algorithms vary in how efficiently they accomplish the task.

The different methods can be compared by counting the number of operations – comparisons and swaps – involved in sorting a given list.

When a computer is used, this translates into the amount of processing time which would be needed.

In Example 1, the bubble sort has involved 14 comparisons and 8 swaps. By using the same list of numbers for each of the methods, you will be able to compare their relative efficiency.

> The 14 comparisons are indicated by the shaded boxes, and the 8 swaps by the double arrows.

Shuttle sort

The **shuttle sort** usually requires fewer comparisons than the bubble sort because it avoids some unnecessary comparisons. It works by sorting the first two members of the list, then the first three, then the first four, and so on.

As with the bubble sort, you can consider the process as a series of passes along the list, as follows.

D1

◆ **First pass**
Compare the first and second numbers. Swap them if they are in the wrong order.

◆ **Second pass**
Compare the second and third numbers. Swap them if they are in the wrong order. If they have to be swapped, compare/swap the first and second numbers.

◆ **Third pass**
Compare the third and fourth numbers. Swap them if they are in the wrong order. If they have to be swapped, compare/swap the second and third numbers. If these were swapped, compare/swap the first and second numbers.

◆ **Subsequent passes**
Repeat the above process, introducing the next number in the list at each stage. Backtrack through the list, comparing and swapping until no swap is necessary.

◆ **Terminating**
The process finishes when the pass which introduces the last number in the list has been completed.

Example 2

Arrange the numbers 4, 8, 2, 6, 3, 5 in ascending order, using the shuttle sort algorithm.

First pass

| 4 | 8 | 2 | 6 | 3 | 5 |

Second pass

4	8 ←→ 2	6	3	5
4 ←→ 2	8	6	3	5
2	4	8	6	3	5

> $8 > 2$, so swap. Compare first and new second numbers:
> $4 > 2$, so swap.

Third pass

| 2 | 4 | 8 ⟷ 6 | 3 | 5 |

| 2 | 4 | 6 | 8 | 3 | 5 |

| 2 | 4 | 6 | 8 | 3 | 5 |

> $8 > 6$, so swap. Compare second and new third numbers.

Fourth pass

| 2 | 4 | 6 | 8 ⟷ 3 | 5 |

| 2 | 4 | 6 ⟷ 3 | 8 | 5 |

| 2 | 4 ⟷ 3 | 6 | 8 | 5 |

| 2 ⟷ 3 | 4 | 6 | 8 | 5 |

| 2 | 3 | 4 | 6 | 8 | 5 |

Fifth pass

| 2 | 3 | 4 | 6 | 8 ⟷ 5 |

| 2 | 3 | 4 | 6 ⟷ 5 | 8 |

| 2 | 3 | 4 ⟷ 5 | 6 | 8 |

| 2 | 3 | 4 | 5 | 6 | 8 |

D1

Efficiency

In Example 2, the shuttle sort has required 12 comparisons and 8 swaps. This is a slight improvement on the bubble sort, which required 14 comparisons. The number of swaps will always be the same in the two methods, as each number can move only one position at a time and has a fixed distance to reach its correct place.

> There are 12 shaded boxes and 8 double arrows.

Shell sort

The **Shell sort**, named after D.L. Shell who published the method in 1959, differs from the two previous methods in that it compares/swaps items which are not next to each other in the list.

The list is divided into a number of sublists, which are separately shuttle sorted. This can happen a number of times, depending on the length of the list to be sorted.

The number of sublists to use at each stage is found by successively dividing the list length by 2, ignoring remainders. This is

indicated by the function $\text{INT}\left(\dfrac{n}{2}\right)$, where n is the length of the list.

The method is illustrated with two examples – the list already used for Examples 1 and 2 and then a longer list.

> For both the bubble and shuttle sorts, the largest number of comparisons needed for a list of n numbers is $\frac{1}{2}n(n-1)$.

> If the list has an even number of numbers, the sublists for the first stage will each contain two numbers.

> If the list has an odd number of numbers, one of the sublists for the first stage will contain three numbers, and the rest two numbers.

Example 3

Arrange the numbers 4, 8, 2, 6, 3, 5 in ascending order, using the Shell sort algorithm.

For the list 4, 8, 2, 6, 3, 5, you have $n = 6$ and so $\text{INT}\left(\dfrac{6}{2}\right) = 3$.

Divide the list into three sublists: the 1st and 4th numbers, the 2nd and 5th, and the 3rd and 6th. These are shuttle sorted. In fact, two of the sublists are already in order.

> The numbers in the sublists are three places apart.

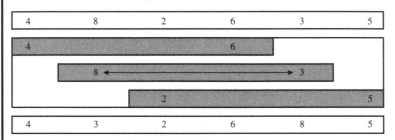

> These are the three sublists.

Then divide into $\text{INT}\left(\dfrac{3}{2}\right) = 1$ sublist, that is, the complete single list. This is then shuttle sorted.

First pass

| 4 ⟷ 3 | 2 | 6 | 8 | 5 |

Second pass

| 3 | 4 ⟷ 2 | 6 | 8 | 5 |

| 3 ⟷ 2 | 4 | 6 | 8 | 5 |

| 2 | 3 | 4 | 6 | 8 | 5 |

Third pass

| 2 | 3 | 4 | 6 | 8 | 5 |

| 2 | 3 | 4 | 6 | 8 | 5 |

Fourth pass

| 2 | 3 | 4 | 6 | 8 | 5 |

| 2 | 3 | 4 | 6 | 8 | 5 |

Fifth pass

| 2 | 3 | 4 | 6 | 8 ⟷ 5 |

| 2 | 3 | 4 | 6 ⟷ 5 | 8 |

| 2 | 3 | 4 | 5 | 6 | 8 |

| 2 | 3 | 4 | 5 | 6 | 8 |

D1

Efficiency

The Shell sort has involved 11 comparisons and 6 swaps in sorting this list, a slight further improvement on the shuttle sort.

> The 11 comparisons are shown by the shaded boxes and the 6 swaps by the double arrows.

Example 4

Sort the list 4, 9, 2, 12, 7, 3, 8, 1, 6 into ascending order, using the Shell sort algorithm.

Here, you have $n = 9$, so $\text{INT}\left(\dfrac{9}{2}\right) = 4$. Divide the list into four sublists: the 1st and 5th and 9th numbers, the 2nd and 6th, the 3rd and 7th, and the 4th and 8th. These are shuttle sorted.

> The numbers in the sublists are four places apart.

D1

| 4 | 9 | 2 | 12 | 7 | 3 | 8 | 1 | 6 |

4				7				6
	9				3			
		2				8		
			12				1	

> These are the four sublists.

| 4 | 3 | 2 | 1 | 6 | 9 | 8 | 12 | 7 |

So far, you have made 6 comparisons and 3 swaps.

Now divide the list into $\text{INT}\left(\dfrac{4}{2}\right) = 2$ sublists, and shuttle sort.

> There are three comparisons – 4, 7; 7, 6 and 4, 6 – in the first sublist and one comparison in each of the other sublists.

| 4 | 3 | 2 | 1 | 6 | 9 | 8 | 12 | 7 |

| 4 | | 2 | | 6 | | 8 | | 7 |
| | 3 | | 1 | | 9 | | 12 | |

| 2 | 1 | 4 | 3 | 6 | 9 | 7 | 12 | 8 |

This has involved a further 8 comparisons and 3 swaps.

Finally, $\text{INT}\left(\dfrac{2}{2}\right) = 1$, so shuttle sort the complete, single list.

| 2 | 1 | 4 | 3 | 6 | 9 | 7 | 12 | 8 |

| 1 | 2 | 3 | 4 | 6 | 7 | 8 | 9 | 12 |

This involved a further 12 comparisons and 5 swaps, making a grand total of 26 comparisons and 11 swaps. Had a simple shuttle sort been done, 27 comparisons and 21 swaps would have been made, while a bubble sort would have involved 36 comparisons and 21 swaps.

> **Note** The use of $\text{INT}\left(\dfrac{n}{2}\right)$ to decide how many sublists to use at each stage is the usual procedure. It would, of course, be possible to use different numbers of sublists. For instance, in the first example, two sublists (4, 2, 3 and 8, 6, 5) could have been used, as the first stage. There has been some research into the efficiency of different schemes, and you might try an investigation of your own, but the use of $\text{INT}\left(\dfrac{n}{2}\right)$ is still the standard approach.

Quicksort

The fourth method you need to study takes a different approach from the comparison/swap method of the other three algorithms (although it is possible to define it in that way – see the note on page 127).

The essence of the method is that one number is chosen as a 'pivot', and the remainder of the list is separated into two sublists: those numbers less than the pivot and those greater. The procedure is then repeated for each sublist, each with its own pivot, and so on until the list is sorted.

As the list starts in random order, the choice of pivot is arbitrary. Some authors use the mid-value from the list, but it is equally valid, and simpler, to use the first item from the list, and that is the approach adopted here.

It is usual to write the sublists with the numbers in the order in which they appeared in the original list. Like the choice of pivot, this is an arbitrary choice, but an algorithm must be definite at each stage, and this convention will be followed here.

The **Quicksort** algorithm is as follows.

> **Step 1** Choose the first number as the pivot.
> **Step 2** For each remaining number in the list:
> ◆ When a number ⩽ pivot, place the number in list A.
> ◆ Otherwise, place the number in list B.
> **Step 3** If any sublist has two or more numbers, apply the Quicksort algorithm from Step 1 for that sublist.

Note This is an example of a **recursive algorithm**. That is, one which is defined in terms of itself. You will meet another example of a recursive algorithm in the solution of the Tower of Hanoi problem in Exercise 9B, Question **3**.

Example 5

Arrange the numbers 4, 8, 2, 6, 3, 5 in ascending order, using the Quicksort algorithm.

| 4 | 8 | 2 | 6 | 3 | 5 |

Taking the 4 as the pivot, you get the following sublists:

| 2 | 3 | 4 | 8 | 6 | 5 |

With the 2 as the pivot for the first sublist, and the 8 for the second, repeat the procedure to obtain:

| 2 | 3 | 4 | 5 | 6 | 8 |

There is now only one sublist with two numbers. Take 5 as the pivot to obtain:

| 2 | 3 | 4 | 5 | 6 | 8 |

The remaining sublists each have length 1, so the process is complete.

Example 6

Arrange the numbers 4, 9, 2, 12, 7, 3, 8, 1, 6 in ascending order using the Quicksort algorithm.

| 4 | 9 | 2 | 12 | 7 | 3 | 8 | 1 | 6 |

Taking the 4 as the pivot, you get the following sublists:

| 2 | 3 | 1 | 4 | 9 | 12 | 7 | 8 | 6 |

With the 2 as the pivot for the first sublist, and the 9 for the second, repeat the procedure to obtain:

| 1 | 2 | 3 | 4 | 7 | 8 | 6 | 9 | 12 |

Finally, with the 7 as the pivot for the only remaining sublist with length above 1, you obtain:

| 1 | 2 | 3 | 4 | 6 | 7 | 8 | 9 | 12 |

and the sort is complete.

D1

In the above layout, the pivot and sublists have been written into a new table, whereas all the previous methods involved only swapping numbers within the same list. However, it is possible to define the Quicksort in a way which just moves items around in the same list, by comparing/swapping the pivot with the unused number at the opposite end of the list, as follows.

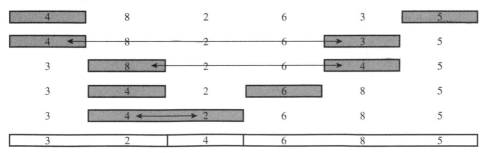

4	8	2	6	3	5

4 < 5, so no swap.
Compare 4 with 3; swap.
Compare 4 with 8; swap.
4 < 6, so no swap.
Compare 4 with 2; swap.

This completes the first iteration of the algorithm. The second iteration would proceed in the same way.

3 ←→ 2		4	6	8	5
2	3	4	6 ← 8 → 5		
2	3	4	5	8 ←→ 6	
2	3	4	5	6	8

First sublist is 3, 2 with 3 as pivot.
Second sublist is 6, 8, 5 with 6 as pivot.

Only remaining sublist is 8, 6.

This completes the sort.

You will notice that this version of the Quicksort may generate the sublists in a different order to the original approach. This does not affect the efficiency of the method.

Efficiency

Defined in this way, the Quicksort algorithm involved 8 comparisons and 6 swaps for this list of numbers, which is the most efficient of the four methods.

> Again, the comparisons are shown by the shaded boxes and the swaps by the double arrows.

Exercise 8A

1 Use the bubble sort algorithm to sort the following list into:

a) ascending order

b) descending order

    12 4 16 5 9 2 4

Show the result of each pass.

D1

2 Repeat question 1, using the shuttle sort algorithm.

3 Sort the following list into ascending order, using the Shell sort algorithm.

    22 26 14 20 12 9 11 15 10

4 Sort the following list into ascending order, using the Quicksort algorithm.

    9 17 6 19 16 13 7 17 12 9

5 A class has students with the following surnames: Harris, Thomas, Patel, Frobisher, Cheung, Allen and Lee.

Sort these into alphabetical order, using a) the bubble sort algorithm and b) the shuttle sort algorithm. Record the number of comparisons and swaps needed for each of the two methods.

6 Repeat question 5 using a) the Shell sort algorithm and (b) the Quicksort algorithm. In each case, record the number of comparisons made.

7 The worst case scenario for a sort is that the original list is in the reverse of the required order.

a) For this worst case, find the number of comparisons and swaps when using i) the bubble sort algorithm and ii) the shuttle sort algorithm on a list of five numbers.

b) How many would be needed, in each case, for a list of i) 10 ii) 20 and iii) n numbers?

8 Compare the Shell sort and the Quicksort algorithms in the worst case scenario with a list of 10 numbers.

Summary

You should know how to ...

1 Apply the **bubble sort** algorithm to an unordered list, as follows:

Work along the list (make a 'pass' through the list) a number of times according to these rules:

First pass
- Compare the first and second numbers. Swap them if they are in the wrong order.
- Successively compare/swap the second and third numbers, the third and fourth numbers, and so on, to the end of the list.

Subsequent passes
- Repeat the above process, leaving out the final number from the previous pass.

Terminating
- The process finishes either when the remaining list has only one number or when a complete pass produces no swaps.

2 Apply the **shuttle sort** algorithm to an unordered list, as follows:

First pass
- Compare the first and second numbers. Swap them if they are in the wrong order.

Second pass
- Compare the second and third numbers. Swap them if they are in the wrong order.
- If they have to be swapped, compare/swap the first and second numbers.

Third pass
- Compare the third and fourth numbers. Swap them if they are in the wrong order.
- If they have to be swapped, compare/swap the second and third numbers.
- If these were swapped, compare/swap the first and second numbers.

Subsequent passes
- Repeat the above process, introducing the next number in the list at each stage. Backtrack through the list, comparing and swapping until no swap is necessary.

Terminating
- The process finishes when the last number has been positioned correctly.

D1

3 Apply the **Shell sort** algorithm to an unordered list, as follows:

For a list of n items:

   Step 1   Divide n by 2, ignoring the remainders, giving m: that is,
   $$m = \text{INT}\left(\frac{n}{2}\right).$$

   Step 2   Divide the list into m sublists. Shuttle sort these sublists.

   Step 3   If $m > 1$, then let $m = \text{INT}\left(\frac{m}{2}\right)$ and go to Step 2.

   Step 4   Stop.

4 Apply the **quicksort** algorithm to an unordered list, as follows:

   Step 1   Choose the first number as the pivot.
   Step 2   For each remaining number in the list:
   - When a number $\leqslant$ pivot, place the number in list A.
   - Otherwise, place the number in list B.
   Step 3   Repeat Steps 1 and 2 until each sublist contains only one number.

5 Compare the efficiency of two methods by counting the number of comparisons and swaps involved in applying them to a given list.

D1

Revision exercise 8

1 There are five mathematicians who are members of a committee:

   Newton (N)   Euler (E)   Descartes (D)   Pythagoras (P)   Archimedes (A)

Use a bubble sort algorithm to rearrange the names into alphabetical order, showing the new arrangement after each comparison. *(AQA, 1999)*

2 a) Use the shuttle sort algorithm to rearrange the following pop groups into alphabetical order, showing the new arrangement after each pass.

   Westlife (W)   Free (F)   Oasis (O)   Cream (C)   U2 (U)
   Abba (A)   Beatles (B)

   b) Find the maximum number of comparisons needed to rearrange a list of 12 pop groups into alphabetical order. *(AQA, 2003)*

3 An array A(N) contains the elements:

   A(1) = 5   A(2) = 2   A(3) = 4   A(4) = 9   A(5) = 1

Use a bubble sort to rearrange the arrays in order of size, showing the new arrangement after each comparison. *(AQA, 1998)*

4 Use the Shell sort algorithm to rearrange the following numbers into ascending order, showing the new arrangement after each pass.

   14 27 23 36 18 25 16 66   (*AQA, 2002*)

5 a) Use the Quicksort algorithm to rearrange the following numbers into order, showing the new arrangement at each stage. Take the first number in any list as the pivot.

   9 5 7 11 2 8 6 17

   b) The shuttle sort algorithm is to be used to rearrange a list of numbers into order.
   i)   Find the maximum number of comparisons that would be needed to be certain that a list of eight numbers was in order.
   ii) Find, in a simplified form, an expression for the maximum number of comparisons that would be needed to be certain that a list containing n numbers was in order.   (*AQA, 1999*) **D1**

6 a) Use the bubble sort algorithm to rearrange the following numbers into ascending order, showing the new arrangement after each pass.

   4 7 13 26 8 15 6 56

   b) Find the maximum number of comparisons needed to rearrange a list of eight numbers into ascending order.   (*AQA, 2001*)

9 Algorithms

This chapter will show you how to ▰▰▰▰▰▰▰▰▰▰

✦ Trace, modify and amend algorithms expressed in words or pseudo-code
✦ Understand that the efficiency of an algorithm depends on the number of operations it requires to arrive at the solution

9.1 Algorithms

D1

In Chapter 1, you met the concept of an algorithm and you have subsequently used a number of algorithms, in relation to network problems and to sorting. You now need to explore algorithms more formally.

An algorithm is a set of instructions for solving a problem. It consists of a sequence of steps which, when followed blindly, lead to the solution – no insight into the problem should be needed. Ideally, an algorithm should also lend itself to implementation on a computer, as most real-world problems are too large to make manual solutions viable.

Definition

An algorithm should have the following properties:

✦ Provide an unambiguous next step at each stage of the solution.
✦ Lead to the solution in a finite number of steps.

The second requirement is important. You need to be certain that you will reach the solution. Also, you should ideally be able to calculate the maximum number of steps (and, therefore, how much processing time) will be required in a particular case.

As an example of a process which fails in this regard, consider the following instructions for writing down the numbers from 1 to 50 in random order.

Step 1 Use the random number function on your calculator to obtain a random number between 1 and 50. (The whole number part of $(1 + 50 \times \text{RAND\#})$ would give this.)
Step 2 Write the number down unless it is already in the list.
Step 3 If the list has 50 numbers, then stop.
Step 4 Go to Step 1.

You can see that as the list grows, the chance of repeats increases. The process has no definite limit to the number of steps required. If you were unlucky, you could go on for a very long time and there is a small probability that the number of steps could exceed any number you care to suggest. The process described is not, therefore, a viable algorithm.

Communicating an algorithm

There are several ways in which an algorithm may be written.

✦ It can be expressed in words.
✦ It can be written in *pseudo-code*. This consists of abbreviated statements using program-like instructions, such as 'repeat ... until', but not using any particular programming language.
✦ It can be drawn as a flowchart. However, this is not included in the present syllabus.
✦ It can be written as a program in a formal computer language. (This will not be required on this course.)

Example 1

Write an algorithm for finding the largest number in a given list.

The basic process is to work along the line, keeping a record of the largest number seen.

This could be expressed more accurately in words as follows.

Step 1 Record the first number in the list.
Step 2 If there are no more numbers in the list, print the recorded number and stop.
Step 3 If the next number is greater than the recorded number, replace the recorded number with it.
Step 4 Go to Step 2

D1

You could write this more formally in pseudo-code. There are several ways of doing this. Here is one possibility.

Use $L(1)$, $L(2)$, $L(3)$, ...$L(n)$ to stand for the 1st, 2nd, 3rd, ... nth numbers in the list. Use the term *count* to stand for the position you have reached, as you move along the list. Use the term *max* to stand for the largest number encountered so far.

A possible list of instructions is:
```
10  INPUT n
20  INPUT L(1) TO L(n)
    .........  [This sets up the list ready for use.]
30  LET count = 1
    .........  [You start in the first position.]
40  LET max = L(count)
    .........  [This records the first number – as in Step 1 above.]
50  LET count = count + 1
    .........  [This moves one step along the list.]
60  IF max < L(count) THEN LET max = L(count)
    .........  [This compares/replaces the recorded number – Step 3.]
70  IF count < n THEN GOTO 50
    .........  [This checks to see whether there are more numbers –
               Step 2.]
80  PRINT max
90  STOP
```

The steps are numbered in increments of 10 to allow later modification with intermediate steps if required.

In the examination, you will not be required to write algorithms from scratch, though you may be asked to modify or amend them.

Tracing an algorithm

When designing or becoming familiar with an algorithm it is often necessary to trace through the process by hand to check what it is doing. It is a good idea to use a *trace table* to record the values of the variables as they are changed by the instructions.

Example 2

Trace the algorithm from Example 1 when applied to finding the largest number in the list L = {25, 30, 32, 17, 50, 49}.

Here, $n = 6$, $L(1) = 25$, $L(2) = 30$, $L(3) = 32$, and so on.

In the trace table, work through the algorithm step by step, and record the values of *max* and *count* as they are changed, as follows.

> You would not usually include a column for notes. It is here to help to explain the process.

max	*count*	L(*count*)	Notes
25	1	25	Record 1st number as *max*.
25	2	30	Compare *max* with 2nd number.
30	2	30	Replace *max*.
30	3	32	Compare *max* with 3rd number.
32	3	32	Replace *max*.
32	4	17	Compare *max* with 4th number.
32	4	17	Do **not** replace.
32	5	50	Compare *max* with 5th number.
50	5	50	Replace *max*.
50	6	49	Compare *max* with 6th number.
50	6	49	Do **not** replace.
Print *max* = 50			*count* = *n*, so end of list.
Stop			Print result and stop.

> You should check that you can follow the table in relation to the pseudo-code instructions in Example 1.

Example 3

Trace the following algorithm, using input values $A = 53$ and $B = 76$. What does the algorithm achieve?

```
10   INPUT A, B
20   LET C = 0
30   IF A IS EVEN THEN GOTO 50
40   LET C = C + B
50   IF A = 1 THEN GOTO 90
60   LET B = B × 2
70   LET A = A ÷ 2 (IGNORING REMAINDERS)
80   GOTO 30
90   PRINT C
100  STOP
```

D1

A	B	C	Notes
53	76	0	Set up starting values.
53	76	76	A is not even, so add B to C.
53	152	76	Double B.
26	152	76	Halve A.
26	304	76	A is even, so C is unchanged. Double B.
13	304	76	Halve A.
13	304	380	A is not even, so add B to C.
13	608	380	Double B.
6	608	380	Halve A.
6	1216	380	A is even, so C is unchanged. Double B.
3	1216	380	Halve A.
3	1216	1596	A is not even, so add B to C.
3	2432	1596	Double B.
1	2432	1596	Halve A.
1	2432	4028	A is not even, so add B to C.
Print 4028 and stop			$A = 1$, so print C and stop.

The result of the algorithm is to print the product $A \times B$.

The algorithm in Example 3 is sometimes known as 'Russian peasant multiplication'. It is a method of multiplying two numbers which requires only knowledge of the two times table and the ability to add. It would be laid out as a pencil and paper calculation as follows.

$$
\begin{array}{rr}
53 & 76 \\
\overline{26} & \overline{152} \\
13 & 304 \\
\overline{6} & \overline{608} \\
3 & 1216 \\
1 & \underline{2432} \\
\text{Product} & = 4028
\end{array}
$$

In the first column, the number is successively divided by 2, ignoring any remainders. In the second column, the number is successively doubled. Any rows with an even number in the first column are crossed out. The second column is then added up to give the answer.

Exercise 9A
. . . .

1 The following pseudo-code listing shows Euclid's algorithm for finding the highest common factor of two numbers. Trace through the algorithm with starting values i) $A = 48$ and $B = 132$; ii) $A = 130$ and $B = 78$.

```
10   INPUT A, B
20   IF A ≤ B THEN GOTO 40
30   SWAP A AND B
40   LET R = REMAINDER FROM B ÷ A
50   IF R = 0 THEN GOTO 90
60   LET B = A
70   LET A = R
80   GOTO 40
90   PRINT 'HCF = ' A
100  STOP
```

2 What would be the output from the following algorithm?

```
10   LET A = 1, B = 1
20   PRINT A, B
30   LET C = A + B
40   PRINT C
50   LET A = B, B = C
60   IF C < 50 GOTO 30
70   STOP
```

3 a) Trace the following algorithm.

```
10   LET A = 1, B = 1
20   PRINT A
30   LET A = A + 2B + 1
40   LET B = B + 1
50   IF B ≤ 10 THEN GOTO 20
60   STOP
```

   b) What would be the effect of the following errors when typing this algorithm?
   i)   Putting 'PRINT B' instead of 'PRINT A'.
   ii) Putting 'LET $A = A + B + 1$' instead of 'LET $A = A + 2B + 1$'.
   iii) Putting '$B < 10$' instead of '$B ≤ 10$'.

4 The algorithm for Russian peasant multiplication shown in Example 3, on page 134, is not as efficient as it could be. As an example, consider what would happen if $A = 127$ and $B = 2$. How would you amend the algorithm to make it more efficient?

5 The following set of instructions forms Zeller's algorithm for finding, from your birth date, the day of the week on which you were born. Apply the algorithm to the date 14 May 1989. (Note: INT means 'the whole number below')

```
10   INPUT day, month, year
20   IF month < 3 THEN LET month = month + 12
     AND LET year = year − 1
30   LET century = INT(year ÷ 100)
40   LET remainder = year − 100 × century
50   LET total = INT(month × 2.6 − 5.39) + INT(century ÷ 4)
     + INT(remainder ÷ 4) + day + remainder − century × 2
60   LET pointer = total − 7 × INT(total ÷ 7)
70   PRINT pointer
```

The algorithm prints a number between and including 0 to 6, which corresponds to a day of the week, with Sunday being 0, Monday being 1, and so on.

6 The following is yet another algorithm for sorting a list of numbers $L(1), L(2), ..., L(n)$ into ascending order.

Use the algorithm to sort the numbers 6, 11, 2, 5, 9, 3, 7, 4 into order, writing down the state of the list after each change.

```
10   INPUT n
20   INPUT L(1) TO L(n)
30   LET position = 1
40   FIND L(min) = MINIMUM OF {L(position+1), ..., L(n)}
50   IF L(position) > L(min) THEN SWAP L(position), L(min)
60   LET position = position + 1
70   IF position < n THEN GOTO 40
80   PRINT L(1) TO L(n)
90   STOP
```

7 Write, in words or pseudo-code, an algorithm for printing the five times table up to twelve fives, using only addition.

8 Write an algorithm, in words or pseudo-code, to detail the steps needed to solve the quadratic equation $ax^2 + bx + c = 0$, starting with the values of a, b and c, and outputting 'no real roots', 'a repeated root $x = ...$' or 'two distinct roots $x = ...$ and $x = ...$'.

9.2 Comparing the efficiency of algorithms

The **efficiency** of an algorithm is measured by calculating how many steps, and, therefore, how much processing time, it requires to find the solution to a problem of a given size. Clearly, if you have a choice of algorithms, the one taking the least time will almost certainly be preferable.

The other consideration when judging the likely usefulness of an algorithm is how rapidly the required number of steps increases as the size of the problem increases. This is called the **order** (or **complexity**) of the algorithm.

For example, the algorithm for finding the largest number in a given list, described in Example 1, on page 133, requires the user or computer to make $(n - 1)$ comparisons for a list of n numbers. (It compares the first or the current 'largest seen yet' with each of the other numbers in the list.) An algorithm like this, where the number of steps is a linear function of the size of the problem, is said to have **linear order**. Typically, for an algorithm of linear order, doubling the size of the problem will approximately double the number of steps needed. For example, a list of 50 numbers would need 49 steps while a list of 100 numbers would need 99 steps.

For some other algorithms, the number of steps grows more rapidly. For example, consider the situation in which a school caretaker has a number of lockers and a bunch of locker keys, but no indication as to which belongs to which. The only systematic algorithm is to try every key in the first locker until the right one is found, then move onto successive lockers with the remaining keys.

The worst-case scenario is that the right key is always the last one tried.

D1

For 10 lockers, the maximum number of tries is

$$10 + 9 + 8 + \ldots + 2 + 1 = 55$$

For n lockers it is

$$n + (n - 1) + (n - 2) + \ldots + 2 + 1 = \tfrac{1}{2}n(n + 1)$$

This is a quadratic function of n and so the algorithm has **quadratic order**.

$\tfrac{1}{2}n(n + 1) = \tfrac{1}{2}n^2 + \tfrac{1}{2}n$ is a quadratic expression.

Typically, for an algorithm of quadratic order, doubling the size of the problem will require approximately four times the number of steps. For example, 20 lockers would need 210 tries.

For some other algorithms, the number of steps grows so rapidly with the size of the problem that the algorithms are not really usable. For example, in the case of the travelling salesman problem of finding the best order in which to visit a number of towns to minimise the travelling distance (Chapter 5, pages 59 to 71), it was not feasible to examine every possible route.

✦ If the salesman needed to visit just six towns, he would have a choice of six first stops, then five next stops, and so on, giving $6! = 720$ possible routes. Each route appears twice, so the number of different routes is 360.

✦ If this rose to 10 towns, he would need to look at $\tfrac{1}{2} \times 10! = 1\,814\,400$ routes.

✦ For n towns, he would have $\tfrac{1}{2} \times n!$ possible routes.

Similarly, suppose you needed to open a combination lock.

✦ With a three-digit combination, you would have to try $10^3 = 1000$ combinations.

✦ For a four-digit combination, you would have to try $10^4 = 10\,000$ combinations.

✦ In general, n digits would correspond to 10^n combinations.

Algorithms in which the number of steps involves $n!$, 10^n, or a similar function are said to have **exponential order**. Algorithms with exponential order are unusable except for small-scale problems.

Note Although you will not be required to answer detailed questions on efficiency and order, you will need to understand, and be able to undertake, the comparison of algorithms in terms of the number of steps involved. You have already seen this when comparing the different sorting algorithms in Chapter 8.

Exercise 9B

1 The following algorithm finds the median of a sorted list of n numbers, labelled L(1), L(2), L(3), ..., L(n).

```
10   INPUT n
20   INPUT L(1), L(2), ..., L(n)
30   LET i = 1, j = n
40   IF j – i < 2 THEN GOTO 70
50   LET i = i + 1 AND j = j – 1
60   GOTO 40
70   LET M = ½(L(i) + L(j))
80   PRINT M
90   STOP
```

a) Follow the above algorithm for these lists.
   i) 2, 3, 3, 7, 8, 8, 12, 14, 15   ii) 9, 9, 12, 14, 19, 19, 22, 23

b) State, with reasons, whether the above algorithm has linear or quadratic order.

2 In Example 1, on page 133 an algorithm is described for finding the largest number in a list and it is noted, on page 137, that a list of n numbers would take $(n - 1)$ comparisons.

This algorithm can be used as the basis for an algorithm for sorting a list of numbers into ascending order. If the list is L(1), L(2), ..., L(n), you proceed as follows:

Step 1   Locate the largest number in the list, and swap it with L(n)
Step 2   Reduce n by 1
Step 3   Repeat Steps 1 and 2 until $n = 1$

a) Use the algorithm given above to sort the list 5, 8, 2, 6, 10, 1, 5, showing the state of the list after each stage. How many comparisons were necessary?

b) How many comparisons will be required for a list of
   i) 10 numbers and ii) 20 numbers.

c) State, with reasons, whether the algorithm has linear or quadratic order.

3 The Tower of Hanoi was a puzzle marketed in 1883 by a French mathematician called Edouard Lucas. It consists of a number of pierced discs of different sizes, which are placed, in decreasing order of size, over one of three pegs. The aim is to move the pile from one peg to another, moving one disc at a time and never placing a larger disc on a smaller one. The solution is illustrated for the case of three discs.

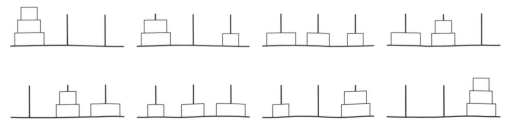

a) As you can see, for three discs, seven moves are required. Investigate the number of moves required for four and five discs. Can you see a pattern?

b) Write an algorithm so that someone could solve the puzzle with any number of discs by following your instructions.

c) The original puzzle was sold with seven discs. How many moves would that take?

d) The puzzle was also called the Tower of Brahma, because of a story (possibly invented by Lucas) that in the city of Benares there is a temple where Brahmin priests move pieces on a puzzle with 64 discs. The legend is that when the puzzle is completed the world will end. Calculate the lifespan of the universe according to this prediction, assuming that they move pieces at the rate of one per second.

e) State, with reasons, the order of the algorithm.

D1

> **Note**   This is another example of a recursive algorithm, similar to the Quicksort algorithm you met in Chapter 8 (page 126). The instructions for moving n discs make reference to moving $(n - 1)$ discs.

Summary

You should know how to ...

1 Write an algorithm in words and in pseudo-code.

2 Trace an algorithm, recording the values of any variables in a trace table.

3 Compare algorithms with each other in terms of their efficiency, relating to the number of steps required to reach each solution.

4 Judge the usefulness of an algorithm in terms of its order or complexity.

Revision exercise 9

1 The algorithm below is used to generate a sequence of numbers.

```
LINE 10   INPUT A, B
LINE 20   PRINT A, B
LINE 30   LET C = A + B
LINE 40   PRINT C
LINE 50   LET A = B
LINE 60   LET B = C
LINE 70   IF C < 10 THEN GOTO LINE 30
LINE 80   END
```

a) Trace the algorithm when $A = 1$ and $B = 1$.

b) Suppose that LINE 70 is changed to:

IF $C \leqslant 50$ THEN GOTO LINE 30

Write down the extra values that C now takes.

c) A student mistypes LINE 60 as LET $C = B$. Find the values of A, B and C that the student would get using the amended algorithm.

(*AQA, 2003*)

2 A student is using the algorithm below to find the real roots of a quadratic equation.

> **Note** For questions **2** and **3**:
> * means multiply.

```
LINE 10    INPUT A, B, C
LINE 20    D = B*B − 4*A*C
LINE 30    X₁ = (−B + √D)/(2*A)
LINE 40    X₂ = (−B − √D)/(2*A)
LINE 50    IF X₁ = X₂ THEN GOTO L
LINE 60    PRINT 'DIFFERENT ROOTS', X₁, X₂
LINE 70    GOTO M
LINE 80    LABEL L
LINE 90    PRINT 'EQUAL ROOTS', X₁
LINE 100   LABEL M
LINE 110   END
```

LINE 10   INPUT A, B, C
LINE 20 $D = B*B − 4*A*C$
LINE 30 $X_1 = (−B + \sqrt{D})/(2*A)$
LINE 40 $X_2 = (−B − \sqrt{D})/(2*A)$
LINE 50   IF $X_1 = X_2$ THEN GOTO L
LINE 60   PRINT 'DIFFERENT ROOTS', X_1, X_2
LINE 70   GOTO M
LINE 80   LABEL L
LINE 90   PRINT 'EQUAL ROOTS', X_1
LINE 100   LABEL M
LINE 110   END

a) Trace the algorithm:
   i)   When $A = 1, B = −4, C = 4$
   ii) When $A = 2, B = 9, C = 9$

b) i)   Find a set of values of A, B and C for which the algorithm would fail.
   ii) Write down additional lines to ensure that the algorithm would not fail for **any** values of A, B and C that may be input.

(*AQA, 2002*)

3 a) Trace the following algorithm.

LINE 1 $A = 1$
LINE 2   LABEL X
LINE 3 $B = A*A*A$
LINE 4   IF $B > 100$ THEN GOTO Y
LINE 5   PRINT A, B
LINE 6 $A = A + 1$
LINE 7   GOTO X
LINE 8   LABEL Y
LINE 9   STOP

b) Explain how your trace table would change if Lines 1 and 2 were interchanged.

(*AQA, 2001*)

D1

4 The following algorithm has been written to input a set of 30 examination marks, each expressed as an integer percentage, and to find the minimum mark and output the result.

```
LINE
10   LET MIN = 100
20   LET I = 0
30   LET I = I + 1
40   INPUT MARK
50   IF MARK < MIN THEN MIN = MARK
60   IF I < 30 THEN GOTO LINE 20
70   .........
80   .........
```

a) i)   State the purpose of Line 20 of the algorithm.

ii) There is a mistake in Line 60. Write down a corrected version of this line.

iii) The contents of Line 70 and Line 80 are missing. Write down the contents of Line 70 and Line 80 to ensure that the algorithm is fully complete.

b) Show how this algorithm could be adapted if the number of examination marks to be input was unknown.

c) Write an algorithm that would input a set of 50 examination marks, each expressed as an integer percentage. Find the maximum mark and output the result.

(AQA, 1999)

5 Some students are experimenting with numbers using the following rules.

'Think of a number;
square it and add the number you first thought of;
square your answer and add the number you first thought of;
continue squaring and adding.'

Here is an algorithm to do this.
```
10        INPUT y
20        LET c = y
30        LET y = y × y + c
40        GOTO 30
```

a) Explain the purpose of the second line '$c = y$'.

b) State what is wrong with this as an algorithm.

c) Write out a modified algorithm which carries out the calculations in the third line exactly 100 times, and which produces an output.

(AQA/NEAB, 1997)

D1 Practice Paper

90 minutes   *75 marks*   *You may use a graphics calculator.*

Answer all questions.

*Note that in the actual examination a write-on table and network will be supplied for use in answering Questions **4** and **7**.*

1   Four pupils, Jon (J), Karen (K), Len (L) and Maisie (M), arrive in the school library to read a newspaper. There are exactly four newspapers available, the *Daily Telegraph* (D), the *Express* (E), the *Financial Times* (F) and the *Guardian* (G). The papers they are prepared to read are shown below.

> Jon will read the *Daily Telegraph* or the *Guardian*.
> Karen will read the *Daily Telegraph* or the *Express*.
> Len will read the *Express* or the *Financial Times*.
> Maisie will read the *Express* or the *Financial Times*.

a) Show this information on a bipartite graph.   *(2 marks)*

b) Initially, Jon picks up the *Daily Telegraph*, Karen the *Express* and Len the *Financial Times*. Demonstrate, by using an alternating path from this initial matching, how a matching that gives everyone a paper they are prepared to read can be achieved.   *(4 marks)*

2   The numbers of parking meters on the roads in a town centre are shown in the network on the right.

A traffic warden wants to start at A, walk along the roads passing each meter at least once, and finish back at A. She wishes to choose her route in order to minimise the number of meters that she passes more than once.

i)   Explain how you know that it will be necessary for her to pass some meters more than once.   *(1 mark)*

ii) Apply the 'Chinese postman' algorithm to find the minimum number of meters which she will have to pass more than once, and give an example of a suitable route.   *(6 marks)*

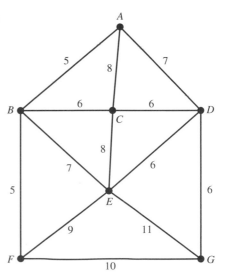

D1

3 a) Apply the following algorithm to the list of data below.

41, 23, 12, 45, 17, 11, 26, 58, 3, 24

Step 1: Choose the first element in the current list; call it P.

Step 2: For each number in the list, put numbers less than P to the left of P, and numbers greater than P to the right of P, creating sublists, each maintaining the original order of numbers.

Step 3: If every sublist has one element, then output the sublists in order. Otherwise go to **Step 1** and repeat for each sublist, keeping the sublists in order.

(5 marks)

b) Find the maximum number of comparisons needed to rearrange a list of ten numbers into ascending order using the algorithm from part a).

(2 marks)

D1

c) Rearrange the list of numbers in part a) so that the maximum number of comparisons would be required.

(1 mark)

4 Some of the eight towns A to H are directly linked by roads. The table shows the distances along these roads in miles.

	A	B	C	D	E	F	G	H
A	–	10	9	5	7	6	7	–
B	10	–	7	8	–	–	8	9
C	9	7	–	–	7	8	–	8
D	5	8	–	–	–	9	9	–
E	7	–	7	–	–	8	9	7
F	6	–	8	9	8	–	6	9
G	7	8	–	9	9	6	–	–
H	–	9	8	–	7	9	–	–

a) In the winter, the local authority grits some of these roads. They wish to grit the minimum total length of roads in order to make it possible to travel between any two of the towns on gritted roads. Use Prim's algorithm starting at A to find which roads they should grit, and state the minimum total length of roads which need to be gritted.

(6 marks)

b) Illustrate in a graph the minimum connector which you found in a). How far is it from A to B in your graph?

(3 marks)

c) The ambulance station is at A and next year the local authority decides that it would like to grit the minimum total length of roads so that the ambulance can reach any of the other towns on gritted roads by travelling less than 15 miles. Show how to adapt your graph in b) to find the roads which the local authority should grit next year.

(3 marks)

5 The Elves toy company makes toy trains and dolls' prams, which use the same wheels and logo stickers.

> Each train requires 8 wheels and 2 logo stickers.
> Each pram requires 8 wheels and 3 logo stickers.
> The company has 7200 wheels and 2200 logo stickers available.
> The company is to make at least 300 of each type of toy and at least 800 toys in total.
> The company sells each train for £20 and each pram for £25.
> The company makes and sells x trains and y prams.
> The company needs to find its minimum and maximum total income, £T.

a) Formulate the company's situation as a linear programming problem.   (*5 marks*)

b) Draw a suitable diagram to enable the problem to be solved graphically, indicating the feasible region and the direction of the objective line.   (*6 marks*)

D1

c) Use your diagram to find the company's minimum and maximum total income, £T.   (*4 marks*)

6 A sweet company has a production line making batches of seven different flavours of sweets A, B, C, D, E, F, G. The changeover times, in minutes, from the production line being set up for one flavour to its being set up for another flavour are given in the following table.

	A	B	C	D	E	F	G
A	–	13	17	18	16	15	14
B	13	–	19	16	18	17.5	16.5
C	17	19	–	24	22	21.5	22
D	18	16	24	–	23	22	21
E	16	18	22	23	–	20	19
F	15	17.5	21.5	22	20	–	18
G	14	16.5	22	21	19	18	–

a) Normally sweets are produced in the order A, B, C, D, E, F, G. The production line is then set up to start the next day with flavour A.
   i) Find the total time taken up by the changeovers.   (*2 marks*)
   ii) Explain why this answer can be considered to be an upper bound for this travelling salesman problem.   (*2 marks*)

b) Use the Nearest Neighbour algorithm, starting at A, to find a reduced time spent on changeovers.   (*4 marks*)

c) By initially ignoring sweet A, find a lower bound for the changeover times.   (*5 marks*)

7 The vertices of the network represent ten towns *A–J*. The arcs
show which towns are linked by direct bus routes, and the numbers
on the arcs show the times of the bus journeys, in minutes.

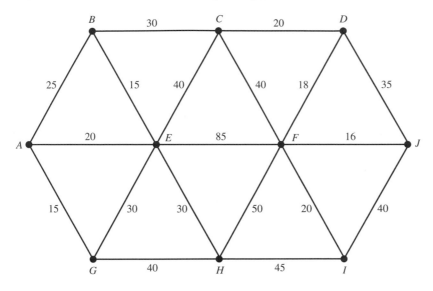

a) i) Use Dijkstra's algorithm on the network shown above to
find the length of the shortest path from *A* to *J*. Show all
your workings at each vertex. *(6 marks)*

ii) Hence, assuming that no time is lost when changing from
one bus to another, state the routes which should be
chosen in order to minimise the time taken when travelling
by bus from *A* to *J*. *(2 marks)*

b) In fact, it is necessary to change buses in each town, and each
change takes 10 minutes. Show that the route chosen in part
a) ii) is no longer the quickest way of getting from *A* to *J* by bus. *(3 marks)*

c) Due to the reduced service on Sundays, each change takes
20 minutes. Find the quickest way of getting from *A* to *J* by
bus on a Sunday. Give a reason for your answer. *(3 marks)*

D1

Answers

Chapter 1
Exercise 1A
1 a) Lane 1: 4, 6, 5, 12, 5, 5; Lane 2: 14, 6, 4, 4, 10; Lane 3: 14, 16, 5, 5; Lane 4: 6, 8, 13, 6, 4; One vehicle of length 4 m unplaced

   b) Lane 1: 12, 14, 14; Lane 2: 16, 10, 13; Lane 3: 4, 6, 5, 5, 5, 6, 4, 4; Lane 4: 5, 6, 5, 8, 6, 4, 4; All vehicles accommodated

2 a) **Step 1** 2 children cross; **Step 2** 1 child returns; **Step 3** 1 adult crosses; **Step 4** The other child returns; **Step 5** If there are more adults, go to Step 1; **Step 6** Stop;   b) 40 times

Exercise 1B
1 (These are examples – others may be possible.)

a)   b)   c)   d)

2 a) (i)   b) (iv)   c) (ii)

Exercise 1C
1 (Your layout may be different.)

a)   b)

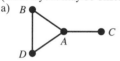

2

	A	B	C	D	E
A	0	1	1	1	1
B	1	0	1	0	1
C	1	1	0	2	0
D	1	0	2	0	1
E	1	1	0	1	0

3

	A	B	C	D	E	F
A	–	14	9	9	–	17
B	14	–	8	–	11	10
C	9	9	–	–	6	–
D	9	–	–	–	12	20
E	–	11	6	12	–	–
F	17	10	–	20	–	–

Chapter 2
Exercise 2A
1 a) CD, AB, BD, BE. Total 22   b) $BC, DE, AB, FE, (GI$ or $HI), (HI$ or $GI), (AH$ or $BH), IF$. Total 307.

2 AB, BD, AC or AB, BD, BC. Both totals 7.

3 a) 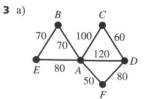   b) $AF, CD, (AB$ or $BE), DF$. Total 330.

4 a) $AF, FG, (GE$ or $FD), (FD$ or $GE), (DB$ or $DC), (DC$ or $DB)$. Total 20.

   b) $AH, (HG$ or $HI), (HI$ or $HG), IC, AB, CD, DE, EF$. Total 178.   **5** FA, AB, BE, FD, DC. Total 330.

6 $AC, CH, AB, BD, DG, GE, EF$. Total 216.

7 a)

	A	B	C	D	E
A	–	5	11	8	6
B	5	–	6	13	4
C	11	6	–	8	8
D	8	13	8	–	10
E	6	4	8	10	–

b) *AB, BE, BC, (AD or CD)*. Total 23.

c) Parts of some roads are used twice. d) Total 21.

8 a)

	O	A	B	C	D	E	F
O	–	4	7	8.5	6.5	4.5	3
A	4	–	3	4.5	4.5	5.5	3
B	7	3	–	2.5	3.5	4.5	4
C	8.5	4.5	2.5	–	2	4	5.5
D	6.5	4.5	3.5	2	–	2	3.5
E	4.5	5.5	4.5	4	2	–	2.5
F	3	3	4	5.5	3.5	2.5	–

b) *OF, FE, ED, DC, CB, (AB or AF)*. Total 15 m.

8 c) Yes, improvement is possible. The author knows of no general algorithm for this problem, but a makeshift strategy gave the following possible solution, length 13.5 m.

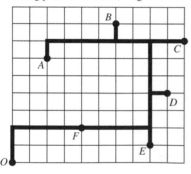

9 a) *EG* (30), *ES* (22), *SI* (20), *SP* (23), *IF* (22)

b)

Total cost £1170

c) Cost £1720. May be better because:
   i) it would take less time if all translations were made in parallel
   ii) slight changes of meaning might build up through a succession of translations.

10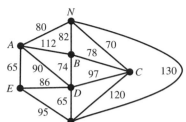

AE, DS, NC, BD, BC, AN
Total cost £432

11 Amend algorithm to choose the heaviest lorry available at each stage. *AG, AB, BC, CH, CD, DI, IF, FE*. 10 tonnes.

12 *ID, IE, IC, CD, FG, GH, CH, AG*, (*AH* or *BH*).

Revision exercise 2

1 a) *AB + BC + (BD or CD) + CE + EF*   b) 33 miles.   c) Does not need to check for cycles.

2 a) Using Kruskal: *AS + TE + TD + (WM + KN* in either order) + (*WA + WK + ME* in any order) + *NR + CR* (total 75 miles). The order of choice when using Prim depends on the starting point.   b) i) Use *KM* for coaches picking up from *M, E, T, D* ii) *K* is more central than *W*.

3 a) *AF + EF + DE + AB + CF*   b)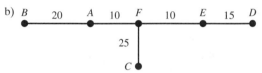

c) Worst cases are *B–C* = 55 min, *B–D* = 55 min, *C–D* = 50 min, all under an hour.

4 a) *AB + AG + GF + FE + ED + GC*, 34 miles.   b) 83 miles.   c) 75 miles.

5 a) 5 vertices need minimum 10 edges to be fully connected.   b) i) 30 units   ii) The edges of weight 7 might form a cycle.
  iii) 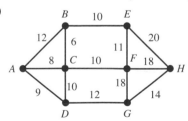   (There are other possible answers.)

6 a) *AC + AE + AD + BC + BF*, 95 miles   b) 2 miles

7 a)   b) *BC + AC + AD + (BE + CF* in either order) + *DG + GH*, £69.
  c) i) Numbers in row *C* halved
    ii) *BC + AC + CF + CD + BE + DG + GH*, £57

Chapter 3

Exercise 3A

1 a)

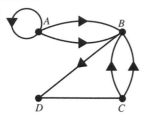

	A	B	C	D
A	0	2	0	0
B	0	1	1	1
C	1	1	2	0
D	1	0	0	0

3 The labels on the vertices should be:
$S\boxed{0}$, $A\boxed{6}$, $B\boxed{9}$, $C\boxed{5}$, D 15 $\boxed{13}$, E 13 $\boxed{12}$, F 19 $\boxed{16}$, G 26 $\boxed{19}$, T 29 28 $\boxed{26}$
The shortest route is $SBEFGT = 26$.

4 The labels on the vertices should be: $S\boxed{0}$, $A\boxed{6}$, $B\boxed{3}$, $C\boxed{8}$, D 16 $\boxed{15}$, $E\boxed{11}$, $T\boxed{19}$

5 a), b)

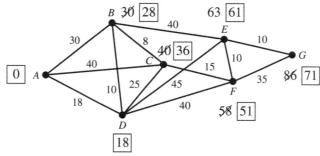

Shortest route $ADBCFEG$, total 71.   c) The table doesn't allow for waiting times.

6 a) A to G: $ACEDG$, total 25. G to A: $GFCEA$, total 21.   **7** Warehouse B. Route $BHJL$, total 25.

8 a) $ABCEF$, total £200.   b) ADF, total £240.   **9** 18 tonnes, route $ABEGH$ or $ACBEGH$.

Revision exercise 3

1 The labels on the vertices should be: $S\boxed{0}$, $W\boxed{5}$, $X\boxed{9}$, T 12 $\boxed{11}$, $U\boxed{17}$, Y 24 $\boxed{16}$, $Z\boxed{23}$, $V\boxed{24}$
The shortest route is $SWTYZ = 23$ km.

2 a) The labels on the vertices should be: $P\boxed{0}$, $R\boxed{8}$, U 16 $\boxed{15}$, $Q\boxed{14}$, $S\boxed{17}$, $T\boxed{19}$, V 27 $\boxed{26}$
The shortest route is $PRTV = 26$ miles.

b) 4 days, $PRSTV$ is only path with edges not more than 10 miles.

3 a) The labels on the vertices should be: $A\boxed{0}$, $B\boxed{30}$, $D\boxed{35}$, E 80 $\boxed{65}$, $C\boxed{75}$, F 95 $\boxed{90}$, G 135 $\boxed{110}$
The shortest route is $ADEFG = 110$ min.

b) $ABFG = 135$ min.

4 a) The labels on the vertices should be: $A\boxed{0}$, $B\boxed{30}$, $E\boxed{45}$, $F\boxed{90}$, $G\boxed{110}$, C 230 $\boxed{210}$, $D\boxed{180}$, $H\boxed{200}$, $I\boxed{240}$
The shortest route is $ABGDI = 240$ p.

b) 79 p   c) i) Prim/Kruskal $AB, AE, EF, BG, GD, CD, DH, HI$ ii) $ABGDHI = 280$ p.

5 a) The labels on the vertices should be: $A\boxed{0}$, $E\boxed{10}$, $H\boxed{15}$, D 30 $\boxed{23}$, K 30 $\boxed{23}$, $B\boxed{30}$, F 38 $\boxed{30}$, G 44 43 $\boxed{38}$, F 53 $\boxed{49}$
The shortest route is $AEHKFGC = 49$ min.

b) 24 min.

6 Working back from N, the labels should be: $N\boxed{0}$, $O\boxed{15}$, $M\boxed{18}$, $L\boxed{25}$, $J\boxed{28}$, K 30 $\boxed{29}$, G 50 $\boxed{41}$, H 69 $\boxed{49}$, I 75 $\boxed{59}$,
$E\boxed{59}$, D 66 $\boxed{64}$, F 79 $\boxed{74}$, $C\boxed{99}$, $B\boxed{100}$, $A\boxed{104}$

b) $ADHGKMN = 104$ s, $BEHGKMN = 100$ s, $CFEHGKMN = 99$ s

7 a) The labels on the vertices should be:

A_0 ⬚0⬚, A_1 ⬚3⬚, A_2 ⬚5⬚, A_3 ⬚4⬚, A_{12} ~~10~~ ⬚9⬚, A_{13} ~~9~~ ⬚8⬚, A_{23} ~~11~~ ⬚10⬚, A_{123} ~~15~~ ~~14~~ ⬚13⬚

The shortest distance = 13.

b) Item 2, item 3, item 1 = £13 000.

8 a) The labels on the vertices should be: A ⬚0⬚, B ⬚4⬚, D ⬚3⬚, F ~~7~~ ⬚6⬚, C ~~11~~ ⬚9⬚, E ~~13~~ ⬚12⬚, G ⬚14⬚, H ~~17~~ ⬚16⬚

ABCEH and *ADFEH*, both with 16 risks.   b) *ABCH* (17 risks).

9 a) The labels on the vertices should be: A ⬚0⬚, B ~~20~~ ~~18~~ ⬚17⬚, G ⬚8⬚, C ⬚12⬚, D ~~42~~ ⬚37⬚, F ~~48~~ ~~47~~ ⬚45⬚, E ~~57~~ ⬚56⬚

Minimum cost £56 million.

b) i) £17 million.   ii) £45 million.   c) $CG + BC + GA + DF + EF + BD$ = £56 million.

Chapter 4

Exercise 4A

1 Repeat *AC* and *CD*. Total length 149. Possible route *ABCDEBDCAECA*.

2 a) Possible pairings *AB/CD* = 14, *AD/BC* = 18, *AC/BD* = 18. Repeat *AE*, *EB*, *CD*. Total length 70.

b) Possible pairings *AB/CF* = 28, *AC/BF* = 29, *AF/BC* = 29. Repeat *AG*, *BG*, *CD*, *DG*, *GF*. Total length 163.

3 a) Repeat *AJI* (or *AKI*) and *HLF* (or *HGF*). Total distance 4040 m.   b) Repeat *IH*. Total distance 3380 m.

4

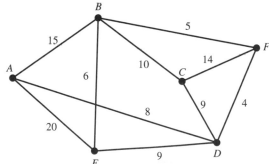

Repeat *AD*, *DC*, *EB*, *BF* or *AD*, *DF*, *EB*, *BC*.
Total distance 128 km.

5 A possible route is *AHCIEFEDCDIFCGHBAGCBA*. This repeats *AB*, *GC*, *CD* and *EF*. Total distance = 880 m.

6 Length 18 m. Double cable on *BAJ*, *IHG* and *DE*, or *BCD*, *EFG* and *IJ*.

Revision exercise 4

1 a) Odd vertices, so not Eulerian   b) *BD* & *FG*, *BF* & *DG*, *BG* & *DF*   c) 5.5, 2.5, 5.5

d) e.g. *ABFBCDGEDEGFECA*, 22 miles.   **2** a) i) e.g. *RWRISCIWCR*.   ii) 1005 m.   b) i) *RW*.   ii) 164.

3 a) i) *GB* & *SW*, *GS* & *BW*, *GW* & *BS*   ii) 94.5 miles   b) 15   c) $(n-1) \times (n-3) \times 3 \times 1$

4 a) 36 units   b) AE, EL, LP   c) 32 units

5 a) Graphs 1 & 2 not Eulerian (odd vertices), Graph 3 Eulerian (all vertices even).

   b) Graph 1: 9 units, Graph 2: 12 units, Graph 3: 12 units.

6 a) i) *CD, DE, EF*   ii) e.g.*ABCDEFEDCEBFA* = 68 km

   a) iii) Network now has 2 directed edges *BC* and *CB* in place of the single edge *BC*. Possible route *ABCDECEBEFBFA* = 77 km.

   b) Start at *F*, end at *C* (or vice versa).

7 a) Odd vertices, so not Eulerian.   b) e.g.*ABCDEBECDA*.   c) e.g.*ABCEBCDA*.

8 a) *A* & *E* are odd vertices.   b) $2.75 < x \leqslant 3$.   c) $211 + 3x$, e.g. *ABCADCEBFEFDEDA*.   d) *ADEF* = $38.5 + x$.

Chapter 5

Exercise 5A

1 a)

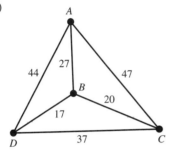

   b) *ABDCA*, total 128.
      Actual route *ABDBCBA*.

2 a) *BEADCB*, total 32, and *EABCDE*, total 29.

   b)

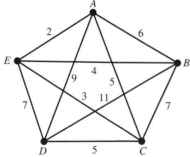

   BECDBA, total 27. Actual route *AECDEBEA*.

3 Best route Hotel – Art Gallery – Cathedral – Museum – Guild Hall – Hotel (or vice versa). Total walking time 24 minutes.

4 a) Frome – Radstock – Shepton Mallet – Glastonbury – Wells – Cheddar – Frome, total 108 km.

   b) Frome – Radstock – Cheddar – Wells – Glastonbury – Shepton Mallet – Frome or vice versa, total 100 km.

5 a)

	A	*B*	*C*	*D*
A	–	5	9	6
B	5	–	10	7
C	9	10	–	3
D	6	7	3	–

   b) *ABDCA*, total 24 miles. Actual route *ABDCDA*.

6 *ADECBA* or vice versa, cost £980.

Exercise 5B

1 a) Edges $(AC + AD) + (BC + CD) = 14$.
Not possible because a tour of this length must include AC, CD and AD, which form a cycle.

b) Edges $(BC + BD) + (AC + AD) = 16$. These edges form a cycle, so optimal tour.

2 a) Best upper bound $420 (starting from C).   b) Best lower bound $360 (deleting A, B or E).

c) £360 ⩽ optimal tour ⩽ $420.

3 a) 251 km   b) $(CD + CL) + (LN + DN + DS) = 169$ km.

4 a)

	A	B	C	D	E
A	–	3	9	6	4
B	3	–	6	7	5
C	9	6	–	9	7
D	6	7	9	–	2
E	4	5	7	2	–

b) Best upper bound = 24.
Best lower bound = 22.

5 a) Best upper bound = 44. Best lower bound = 44. So optimal route $SBACEDS$ = 44 points.

b) 81 ⩽ best route ⩽ 84. Best route from $SCBDAES$ = 81 points.

6 a)

	A	B	C	D	E	F
A	–	4	2	4	3	5
B	4	–	2	4	3	5
C	2	2	–	4	3	3
D	4	4	4	–	3	3
E	3	3	3	3	–	2
F	5	5	3	3	2	–

b) Upper bound 16 moves.
Lower bound 15 moves.

Revision exercise 5

1 a) $PSQTURVP$, total 34 miles.   b)   25 miles.

c) A lower bound is $25 + (3 + 4) = 32$. An upper bound is 34 from part a). Hence, $32 < L < 34$.

2 a) $DFGEJIHD$, 53 min.   b) EG, EJ, HI, IJ, FG, 40 min, $F – G – E – J – I – H$.

c) A lower bound is $40 + (6 + 7) = 53$ min, so $DFGEJIHD$ is optimal.

3 a) i) ACBDA = 31 min.   ii) 6.   b) $n!$.

4 a) 39.   b) $SRQPTVUS$ = 16.   c) i) Cycle must use two edges at U, one of which must be 3.

c) ii)   The 'bottleneck' at S would prevent a Hamiltonian cycle if PT and RV could not be used.

iii)   Cycle must be at least $4 + 3 + 1 + 2 + 2 + 2 + 2 = 16$.   d) i) 136 s.   ii) 339 s.

5 a) $LOVRMCL$ = 175, $OVRMCLO$ = 175, $COVRMLC$ = 172, $MROVLCM$ = 175.   b) $OVLRMCO$ = 168 min.

6 a) $x < 9, x < 11, x > 3$.   b) $x > 7, x > 3$.   c) $x = 8$, tour = 104 space units.

7 a)

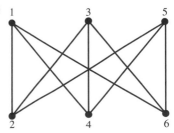

b) i e.g. 1234561.   ii) Even $n > 2$.
c) $\frac{1}{2}n$.
d) If n is even, need $\frac{1}{2}n$ to be even, so n must be a multiple of 4.
If n is odd, there will be odd vertices, so cannot be Eulerian.

Chapter 6

Exercise 6A

1 a)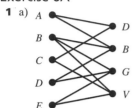

2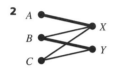

3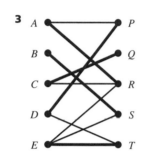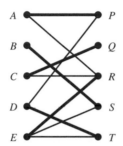

4 a)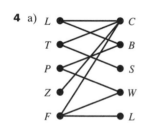

b) Matching *FL*, *ZC*, *LB*, *TS*, *PW*.
All choices forced, so only possible matching.

5 a)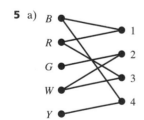

b) e.g. *B*1, *G*2, *R*3, *Y*4.

Exercise 6B

1 a) $A1, B5, C2, D3, E4$. b) Maximal. c) Alternating path $D4A2$. Max. matching $A2, B1, C3, D4$.

d) Alternating paths $C1B3$ and $F6A2E5$. Max. matching $A2, B3, C1, D4, E5, F6$.

e) Alternating path $D3E5$. Max. matching $B4, C1, D3, E5$.

2 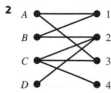 Alternating path $A3C4$.
Complete matching $A3, B1, C4, D2$.

3 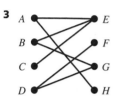 Alternating path $CEAH$.
Complete matching AH, BG, CE, DF.

4 a), b) 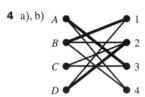 c) Alternating path $C3A4$, matching $A4, B1, C3, D2$.
Alternating path $C2D1B4$, matching $A3, B4, C2, D1$.

5 a) 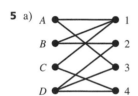 Possible matchings $A3, B2, C1, D4$ or $A1, B2, C4, D3$ or $A3, B1, C4, D2$.
b) Cheapest is $A3, B2, C1, D4$, cost £980.

Revision exercise 6

1 a) 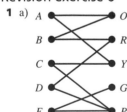 b) Alternating path $BRDPEG$, matching AO, BR, CY, DP, EG.

2 a) 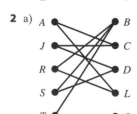 b) Alternating path $JCALRBTO$, matching AL, JC, RB, SD, TO.

3 a)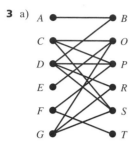

b) Possible alternating paths *ABDR* and *EOCPFT*, giving matching *AB*, *CP*, *DR*, *EO*, *FT*, *GS*. There are other possible alternating paths leading to matchings *AB*, *CP*, *DS*, *EO*, *FT*, *GR* or *AB*, *CS*, *DP*, *EO*, *FT*, *GR*.

4 a)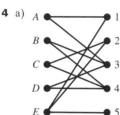

b) Possible alternating paths *C2* and *D3A1E5* giving matching *A1*, *B4*, *C2*, *D3*, *E5*. There are other possible alternating paths and a second possible matching *A1*, *B3*, *C2*, *D4*, *E5*.

Chapter 7

Exercise 7A

(These inequalities have not been simplified.)

1 x litres of Econofruit, y litres of Healthifruit.
Maximise $30x + 40y$ subject to $0.2x + 0.4y \leq 20\,000$, $0.5x + 0.3y \leq 30\,000$, $x \geq 0$, $y \geq 0$.

2 x cars, y minibuses. Minimise $20x + 60y$ subject to $5x + 12y \geq 80$, $x + y \leq 8$, $x \geq 0$, $y \geq 0$, x and y are integers.

3 x ha of wheat, y ha of potatoes. Maximise $80x + 100y$ subject to $30x + 50y \leq 2800$, $700x + 400y \leq 40\,000$, $x + y \leq 75$, $x \geq 0$, $y \geq 0$.

4 x oranges, y apples, z pears. Minimise $20x + 12y + 15z$ subject to $x \leq y$, $y \geq 2z$, $x + y + z \geq 30$, $x > 0$, $y > 0$, $z > 0$, x, y, z are integers.

5 x litres of whisky, y litres of ginger wine. Minimise $12x + 5y$ subject to $x + y \geq 0.1$, $0.4x + 0.12y \leq 0.03$, $0.4x + 0.12y \geq 0.2 (x + y)$, $x > 0$, $y > 0$.

Exercise 7B

1 a)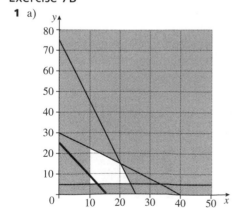

The graph shows the objective line at $P = 50$.
Optimal value $P = 90, x = 20, y = 15$

b)

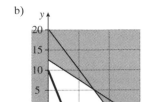

The graph shows the objective line at $P = 8$.
Optimal value $P = 32, x = 8, y = 0$.

c)

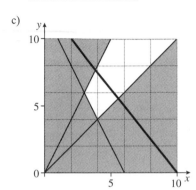

The graph shows the objective line at $C = 50$.
Optimal value $C = 39, x = 3, y = 6$.

d)

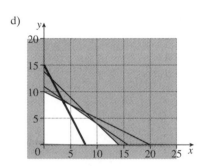

The graph shows the objective line at $R = 15$.
Optimal value $R = 28, x = 14, y = 0$.

2

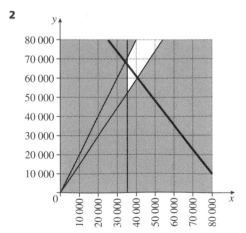

Objective line drawn at $C = 2.2$ million.
Optimal $x = 35\,000,\ y = 52\,500$.

3

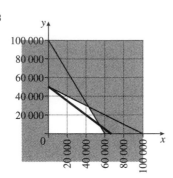

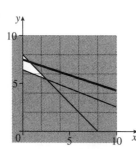

Objective line drawn at $P = 2\,000\,000$.
Optimal value $P = 2\,428\,571$, $x = 42\,857$, $y = 28\,571$.

4

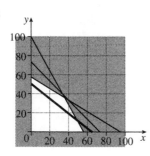

Objective line drawn at $C = 450$.
Optimal value $C = 400$, $x = 2$, $y = 6$.

5

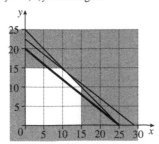

Objective line drawn at $P = 5000$.
Optimal value $P = 6340$, $x = 38$, $y = 33$.
(Note that the constraint $x + y \leqslant 75$ is redundant.)

6 x houses, y bungalows. Maximise $P = 20\,000x + 25\,000y$ subject to $x + y \leqslant 25$, $210x + 270y \leqslant 6000$, $x \leqslant 15$, $y \leqslant 15$, $x \geqslant 0$, $y \geqslant 0$, x, y are integers.

Objective line drawn at $P = 500\,000$.
Optimal value $P = 560\,000$, $x = 13$, $y = 12$.

7 a) Minimise $C = 20x + 15y + 12z$ subject to $x + y + z = 50, 2x + 2y \geqslant z, x \geqslant y, x \geqslant 10, y \geqslant 5, x, y, z \geqslant 0, x, y, z$ are integers.

c)

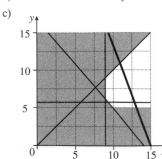

Objective function drawn at $C = 720$.
Optimal value $C = 701, x = 10, y = 7, z = 33$.

Revision exercise 7

1 b) i)

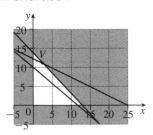

ii) Objective line drawn at $P = 10x + 12y = 120$.
   Vertex V shown on graph.
iii) Intersect at non-integer point.
iv) $x = 2$, $y = 11$ giving $P = £152$.

2 a) $3x + 2y + z \leqslant 120, 3x + 4y + 6z \leqslant 330$.

b) ii), iii) Objective line drawn at $T = x + 2y = 20$.   iv) Maximum $T = 70$.   v) Maximum $T = 64$.

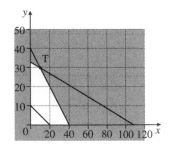

3 a) i) 400.   ii) 50.   b) $x \geqslant 0, y \geqslant 0, x + y \geqslant 50, 2x + y \leqslant 200, 2x + 5y \leqslant 600$.

4 a) $2 \leqslant x \leqslant y, 6 \leqslant x + y \leqslant 10$.

b)

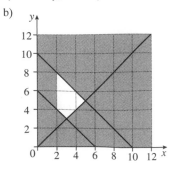

c) Minimum cost £18, maximum cost £36.

d) £18, £20, £22, ..., £36.

5 a) $(1000 - x - y)$.

b)

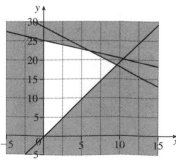

c) ii) Minimum cost £37 500, sending 500 by road and 500 by sea.　iii) 33.

6 a) $x + y + z \leqslant 11$, $3x + y + 2z \leqslant 22$.　b) $P = 4x + 3y + z$　c) $3x + y \leqslant 20$.

d)

Maximum profit £36, when $x = y = 5$.　e) Maximum profit £38, when $x = 5, y = 6$.

7 a) $\frac{1}{2}x + 6y$ is the weight of strawberries for x jam and y jelly.　$x + 7y \leqslant 60$, $x + 2y \leqslant 40$.

b)

c) 104.　d) 50 of each.

8 a) x = no. of sets of pans, y = no. of sets of crockery:
$y \geqslant 2x$, $x + y \leqslant 28$, $x + 2y \leqslant 50$.

b)

c) Maximum profit £78 when $x = 6$, $y = 22$.

d) $x = 9$, $y = 19$.

9 a) $c \geqslant 20$, $d \geqslant 20$, $e \geqslant 20$, $2c + 2d + 3e \leqslant 200$, $e < c + d$.　c) $c = d = 20$.

Chapter 8

Exercise 8A

1 a)
```
12   4  16   5   9   2   4
 4  12   5   9   2   4  16
 4   5   9   2   4  12  16
 4   5   2   4   9  12  16
 4   2   4   5   9  12  16
 2   4   4   5   9  12  16
 2   4   4   5   9  12  16
```
b)
```
12   4  16   5   9   2   4
12  16   5   9   4   4   2
16  12   9   5   4   4   2
16  12   9   5   4   4   2
```

2 a)
```
12   4  16   5   9   2   4
 4  12  16   5   9   2   4
 4  12  16   5   9   2   4
 4   5  12  16   9   2   4
 4   5   9  12  16   2   4
 2   4   5   9  12  16   4
 2   4   4   5   9  12  16
```
b)
```
12   4  16   5   9   2   4
12   4  16   5   9   2   4
16  12   4   5   9   2   4
16  12   5   4   9   2   4
16  12   9   5   4   2   4
16  12   9   5   4   2   4
16  12   9   5   4   4   2
```

3
```
22  26  14  20  12   9  11  15  10
10   9  11  15  12  26  14  20  22
10   9  11  15  12  20  14  26  22
 9  10  11  12  14  15  20  22  26
```

4
```
6  7  9  9  17  19  16  13  17  12
6  7  9  9  16  13  17  12  17  19
6  7  9  9  13  12  16  17  17  19
6  7  9  9  13  13  16  17  17  19
```

5 a)

							Comp	Swap
Harris	Thomas	Patel	Frobisher	Cheung	Allen	Lee		
Harris	Patel	Frobisher	Cheung	Allen	Lee	Thomas	6	5
Harris	Frobisher	Cheung	Allen	Lee	Patel	Thomas	5	4
Frobisher	Cheung	Allen	Harris	Lee	Patel	Thomas	4	3
Cheung	Allen	Frobisher	Harris	Lee	Patel	Thomas	3	2
Allen	Cheung	Frobisher	Harris	Lee	Patel	Thomas	2	1
Allen	Cheung	Frobisher	Harris	Lee	Patel	Thomas	1	0
						Total	21	15

b)

							Comp	Swap
Harris	Thomas	Patel	Frobisher	Cheung	Allen	Lee		
Harris	Thomas	Patel	Frobisher	Cheung	Allen	Lee	1	0
Harris	Patel	Thomas	Frobisher	Cheung	Allen	Lee	2	1
Frobisher	Harris	Patel	Thomas	Cheung	Allen	Lee	3	3
Cheung	Frobisher	Harris	Patel	Thomas	Allen	Lee	4	4
Allen	Cheung	Frobisher	Harris	Patel	Thomas	Lee	5	5
Allen	Cheung	Frobisher	Harris	Lee	Patel	Thomas	3	2
						Total	18	15

6 a)

							Comp
Harris	Thomas	Patel	Frobisher	Cheung	Allen	Lee	
Frobisher	Cheung	Allen	Harris	Thomas	Patel	Lee	4
Allen	Cheung	Frobisher	Harris	Lee	Patel	Thomas	9
						Total	13

b)

							Comp
Harris	Thomas	Patel	Frobisher	Cheung	Allen	Lee	
Frobisher	Cheung	Allen	Harris	Thomas	Patel	Lee	6
Cheung	Allen	Frobisher	Harris	Patel	Lee	Thomas	4
Allen	Cheung	Frobisher	Harris	Lee	Patel	Thomas	2
						Total	12

7 a) Both require 10 comparisons and 10 swaps.   b) i) Both require 45 comparisons and 45 swaps.

ii) Both require 190 comparisons and 190 swaps.   iii) Both require $\frac{1}{2}n(n-1)$ comparisons and $\frac{1}{2}n(n-1)$ swaps.

8 Shell sort involves 27 comparisons and 13 swaps, Quicksort 45 comparisons (and 5 swaps if you use the comparison/swap version of Quicksort).

Revision exercise 8

1

N	E	D	P	A
E	N	D	P	A
E	D	N	P	A
E	D	N	A	P
D	E	A	N	P
D	A	E	N	P
A	D	E	N	P

2 a)

W	F	O	C	U	A	B
F	W	O	C	U	A	B
F	O	W	C	U	A	B
C	F	O	W	U	A	B
C	F	O	U	W	A	B
A	C	F	O	U	W	B
A	B	C	F	O	U	W

b) 66

3

5	2	4	9	1
2	5	4	9	1
2	4	5	9	1
2	4	5	1	9
2	4	1	5	9
2	1	4	5	9
1	2	4	5	9

4

14	27	23	36	18	25	16	66
14	25	16	36	18	27	23	66
14	25	16	27	18	36	23	66
14	16	18	23	25	27	36	66

5 a)

9*	5	7	11	2	8	6	17
5*	7	2	8	6	9	11*	17
2*	5	7*	8	6	9	11	17*
2	5	6*	7	8*	9	11	17
2	5	6	7	8	9	11	17

5 b) i) 28   ii) $\frac{1}{2}n(n-1)$

6 a)

4	7	13	26	8	15	6	56
4	7	13	8	15	6	26	56
4	7	8	13	6	15	26	56
4	7	8	6	13	15	26	56
4	7	6	8	13	15	26	56
4	6	7	8	13	15	26	56

b) 28

Chapter 9
Exercise 9A

1 i)

A	48	36	12
B	132	48	36
R	36	12	0

HCF = 12

ii)

A	130	78	52	26
B	78	130	78	52
R	–	52	26	0

HCF = 26

2 1, 1, 2, 3, 5, 8, 13, 21, 34, 55 (the Fibonacci series)

3 a)

A	1	4	9	16	25	36	49	64	81	100	121
B	1	2	3	4	5	6	7	8	9	10	11

Printout: 1, 4, 9, 16, 25, 36, 49, 64, 81, 100

b) i) Print: 1, 2, 3, 4, 5, 6, 7, 8, 9, 10   ii) Print: 1, 3, 6, 10, 15, 21, 28, 36, 45, 55   iii) Print: 1, 4, 9, 16, 25, 36, 49, 81

4 Insert a line 15 IF $A > B$ THEN SWAP A AND B **5** Sunday

6

2	11	6	5	9	3	7	4
2	3	6	5	9	11	7	4
2	3	4	5	9	11	7	6
2	3	4	5	9	11	7	6
2	3	4	5	6	11	7	9
2	3	4	5	6	9	7	11
2	3	4	5	6	7	9	11
2	3	4	5	6	7	9	11

7 Here is a suggested algorithm:

10	LET $A = 1, B = 5$
20	PRINT A '× 5 =', B
30	LET $A = A + 1$
40	LET $B = B + 5$
50	If $A < 13$ GOTO 20
60	STOP

8 Here is a suggested algorithm:
```
10   INPUT A, B, C
20   LET D = B² − 4AC
30   IF D < 0 THEN PRINT 'NO REAL ROOTS' AND STOP
40   IF D = 0 THEN GO TO 90
50   LET R₁ = (−B + √D)/2A
60   LET R₂ = (−B − √D)/2A
70   PRINT 'TWO DISTINCT ROOTS', R₁, R₂
80   STOP
90   LET R = −B/2A
100  PRINT 'REPEATED ROOT', R
110  STOP
```

Exercise 9B

1 a) i) 8.
   ii) 16.5.

b) The number of steps $= \text{INT}(\frac{1}{2}(n + 1))$, so it is of linear order.

2 a)

							Comp
5	8	2	6	10	1	5	
5	8	2	6	5	1	10	6
5	1	2	6	5	8	10	5
5	1	2	5	6	8	10	4
5	1	2	5	6	8	10	3
2	1	5	5	6	8	10	2
1	2	5	5	6	8	10	1

Total comparisons 21

b) i) 45.   ii) 190.   c) Doubling the problem size gives approx. $4 \times$ no. of steps, so quadratic order (in fact, $\frac{1}{2}n(n - 1)$ steps)

3 a) 3 discs, 7 moves: 4 discs, 15 moves: 5 discs, 31 moves ... n discs, $2^n - 1$ moves.

b) Let the pegs be A, B and C. To move n discs from A to B: **Step 1** Move the top $(n - 1)$ discs from A to C; **Step 2** Move the remaining disc from A to B; **Step 3** Move the $(n - 1)$ discs from C to B.
This is a recursive algorithm. To move 2 discs (Move 2), you use the algorithm with $n = 2$. To move 3 discs (Move 3), you use Move 2 at Steps 1 and 3 (with appropriate letters). For Move 4, you use Move 3 at Steps 1 and 3. And so on.

c) $2^7 - 1 = 127$ moves   d) 1.845×10^{19} seconds = approx. 585 billion years   e) Exponential, because it involves 2^n.

Revision exercise 9

1 a)

A	B	C
1	1	2
1	2	3
2	3	5
3	5	8
5	8	13
8	13	

b) 21, 34, 55   c)

A	B	C
1	1	2
1	1	1
1	1	2
1	1	1
Repeats indefinitely		

Print: 1, 1, 2, 3, 5, 8, ...   Print: 2, 2, 2, 2, ...

2 a) i)

A	B	C	D	X_1	X_2
1	−4	4			
			0		
				2	2

ii)

A	B	C	D	X_1	X_2
2	9	9			
			9		
				−1.5	−3

b) i)   Any set of values where $D < 0$ or $A = 0$

ii)   Line 15 IF $A = 0$ THEN PRINT 'NOT QUADRATIC': GO TO M
Line 25 IF $D < 0$ THEN PRINT 'NO SOLUTIONS': GO TO M

3 a)

A	B
1	1
2	8
3	27
4	64
5	125

b) A always equals 1 and algorithm loops endlessly.

4 a) i) Setting the counter to zero initially   ii) IF I < 30 THEN GOTO LINE 30   iii) 70 PRINT MIN, 80 STOP

b) Remove lines 20, 30 and 60.
Insert Line 45 IF MARK > 100 THEN GOTO LINE 80
Use 101 as the final input value.

c) Original algorithm with the following changes:
10 LET MAX = 0
50 IF MARK > MAX THEN MAX = MARK
60 IF I < 50 THEN GOTO LINE 30
70 PRINT MAX
80 STOP

5 a) To keep a copy of the original number.   b) No output, and never terminates.

c)
10 INPUT y
20 LET $c = y$
30 LET I = 0
40 LET I = I + 1
50 LET $y = y \times y + c$
60 IF I < 100 THEN GOTO LINE 40
70 PRINT y
80 STOP

D1 Practice Paper

1 a)
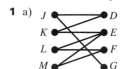

Alternating path *MEKDJG*.
Matching *JG*, *KD*, *LF*, *ME*.
(Other solutions possible.)

2 i) There are odd vertices, so not Eulerian.

ii) Odd vertices *A, E, F, G*.
Possible pairings: $AE + FG = 22, AF + EG = 21, AG + EF = 22$.
Best route repeating *AF* and *EG* = 94 + 21 = 115.
Possible route *ABFGEFBADGEBCDECA*.

3 a)

41*	23	12	45	17	11	26	58	3	24
23*	12	17	11	26	3	24	41	45*	58
12*	17	11	3	23	26*	24	41	45	58*
11*	3	12	17*	23	24*	26	41	45	58
3*	11	12	17	23	24	26	41	45	58

b) 45   c) 58, 45, 41, 26, 24, 23, 17, 12, 11, 3

4 a) *AD* (5), *AF* (6), *FG* (6), *AE* (7), *EC* (7), *EH* (7), *BC* (7), total 45. (You may have chosen the 7s in a different order.)

b)
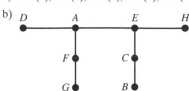

From *A* to *B* is *AECB* = 21

4 c) *AB* is too long. Need to grit *BD* instead of *BC* (adding one extra mile of gritting).

5 a) Objective function $T = 20x + 25y$. Constraints $x + y \leqslant 900, 2x + 3y \leqslant 2200, x \geqslant 300, y \geqslant 300, x + y \geqslant 800$.

b)

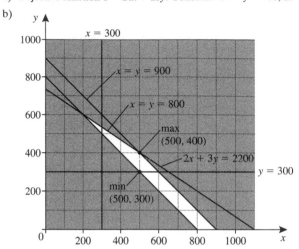

c) Minimum $T = 17\,500$ at $(500, 300)$, maximum $T = 20\,000$ at $(500, 400)$.

6 a) i) 131 mins.   ii) It is a possible solution, but might be improved.

b) $ABDGFECA = 127$.   c) Minimum connector BD, BG, BF, BE, BC, total 87. Lower bound $= 87 + (13 + 14) = 114$.

7 a) i) Labels on vertices should be:

$A \boxed{0}$, $G \boxed{15}$, $E \boxed{20}$, $B \boxed{25}$, H ~~55~~ $\boxed{50}$, C ~~60~~ $\boxed{55}$

$D \boxed{75}$, F ~~105 100 95~~ $\boxed{93}$, $I \boxed{95}$, J ~~110~~ $\boxed{109}$

Length of shortest path $= 109$ min.

ii) Route is $ABCDFJ$.

b) $ABCFJ$ and $AEFJ$ take 141 minutes with changes, better than $ABCDFJ$, which takes 149 minutes with changes.

c) Fewest changes by travelling $AEFJ = 121 + 2 \times 20 = 161$. Any other route has at least three changes and so takes at least $3 \times 20 +$ shortest route (109) $= 169$ mins.

Index